Fundamentals of Switching Theory and Logic Design

Fundamentals of Switching Theory and Logic Design

Duncan Hodgson

Larsen & Keller
www.larsen-keller.com

Fundamentals of Switching Theory and Logic Design
Duncan Hodgson
ISBN: 979-8-88836-043-9 (Hardback)

Larsen & Keller
Published by Larsen and Keller Education,
5 Penn Plaza,
19th Floor,
New York, NY 10001, USA

Cataloging-in-Publication Data

Fundamentals of switching theory and logic design / Duncan Hodgson.
p. cm.
Includes bibliographical references and index.
ISBN 979-8-88836-043-9
1. Switching theory. 2. Logic design. 3. Logic circuits. I. Hodgson, Duncan.
TK7868.S9 F86 2023
621.381 537--dc23

For more information regarding Larsen and Keller Education and its products, please visit the publisher's website www.larsen-keller.com

Table of Contents

Preface

This book has been an outcome of determined endeavour from a group of educationists in the field. The primary objective was to involve a broad spectrum of professionals from diverse cultural background involved in the field for developing new researches. The book not only targets students but also scholars pursuing higher research for further enhancement of the theoretical and practical applications of the subject.

Switching theory refers to the abstract mathematical formalization used in the logic design of digital networks. The modern logic networks are usually constructed using electronic integrated circuits (ICs) which consist of logical elements such as inverters, AND gates, and OR gates. The basic component of modern digital system is two-valued Boolean algebra, sometimes referred to as switching algebra. It is used as a notation to represent the operation of such logic networks. Boolean algebra performs three basic operations. These include inversion, logical addition, and logical multiplication. These operations are implemented directly by logic gates called inverters, OR gates, and AND gates. Switching theory plays an instrumental role in data processing, telephone, and other technologies wherein the decisions are made about routing information. This book provides in-depth knowledge of switching theory and logic design. Researchers and students in the field of electronics and communication, and electrical engineering will be greatly assisted by it.

It was an honour to edit such a profound book and also a challenging task to compile and examine all the relevant data for accuracy and originality. I wish to acknowledge the efforts of the contributors for submitting such brilliant and diverse chapters in the field and for endlessly working for the completion of the book. Last, but not the least; I thank my family for being a constant source of support in all my research endeavours.

Duncan Hodgson

1

Review of Number Systems and Codes

1.1 Representation of Numbers of Different Radix and Conversation of Radix

Binary Numbers

- The decimal number say 7392 can be represented as:

  $7 \times 10^3 + 3 \times 10^2 \times 9 \times 10^1 \times 2 \times 10^0$

- A number with a decimal point can be represented by a series of coefficients as follows:

  $a_5 a_4 a_3 a_2 a_1 a_0 . a_{-1} a_{-2} a_{-3}$

- The decimal equivalent for the binary 11010.11 is 26.75 which can be represented as:

  $1 \times 2^4 + 1 \times 2^3 + 0 \times 2^2 + 1 \times 2^1 + 0 \times 2^0 + 1 \times 2^{-1} + 1 \times 2^{-2} = 26.75$

- A number expressed in base-*r* system will have coefficients that are multiplied by powers of *r*:

  $a_n . r^n + a_{n-1} . r^{n-1} + \ldots + a_2 . r^2 + a_1 . r + a_0 + a_{-1} . r^{-1} + a_{-2} . r^{-2} + \ldots + a_{-m} . r^{-m}$

Here, r = 2, 3, 4 ... 8, 9, 10, ... 16 ...

System	Radix	Allowable Digits
Hexadecimal	16	0,1,2,3,4,5,6,7,8,9,A, B, C, D, E, F
Decimal	10	0,1,2,3,4,5,6,7,8,9
Octal	8	0,1,2,3,4,5,6,7
Binary	2	0,1

Examples:

1. $(4021.2)_5 = 4 \times 5^3 + 0 \times 5^2 + 2 \times 5^1 + 1 \times 5^0 + 2 \times 5^{-1}$

$= 4 \text{ X } 125 + 0 + 10 + 1 + 2 \text{ X } (1/5)$

$= 500 + 11 + .4$

$= (511.4)_{10}$

2. $(B65F)_{16} = 11 \times 16^3 + 6 \times 16^2 + 5 \times 16^1 + 15 \times 16$

$= 11 \times 4096 + 6 \times 256 + 5 \times 16 + 15$

$= 45056 + 1536 + 80 + 15$

$= (46687)_{10}$

Augend	101101	Minuend	101101	multiplicand	1011
Addend: +	100111	Subtrahend: -	100111	Multiplier:	×101
	1010100		000110		110111

Number Base Conversions

Decimal

- A binary number is converted to a decimal by forming the sum of powers of 2 of those coefficients whose value is 1.

Example:

$$(1010.011)_2 = 2^3 + 2^1 + 2^{-2} + 2^{-3} = (10.375)_{10}$$

- Thus, a number expressed in base r can be converted to its decimal equivalent where it is multiplied with each coefficient of the corresponding power of r and then adding it gives the result.

Example:

$$(630.4)_8 = 6\times8^2 + 3\times8^1 + 0\times8^0 \times 4\times8^{-1} = (408.5)_{10}$$

Conversion from Decimal Fraction $(0.6875)_{10}$ to Binary

Example:

The conversion from a decimal fraction to any base-r system is similar to the example shown above. The multiplication is done by r instead of 2, therefore the coefficients found from the integers may range in value from 0 to r-1 instead of 0 and 1.

Coefficients	Integer	Fraction
0.6875 × 2 = 1	+ 0.3750	$a_{-1} = 1$
0.3750 × 2 = 0	+ 0.7500	$a_{-2} = 0$
0.7500 × 2 = 1	+ 0.5000	$a_{-3} = 1$
0.5000 × 2 = 1	+ 0.0000	$a_{-4} = 1$

Conversion from Decimal 41 to Binary

Example:

Remainder Coefficient	Integer	Quotient
41/2 = 20	+ ½	$a_0 = 1$
20/2 = 10	0	$a_1 = 0$
10/2 = 5	0	$a_2 = 0$
5/2 = 2	+ ½	$a_3 = 1$
2/2 = 1	0	$a_4 = 0$
1/2 = 0	+ ½	$a_5 = 1$

The conversion from the decimal integers to any base-r system is similar to the example given above, except that division is done by r instead of 2.

Conversion from Decimal 153 to Octal

Example:

153				
19	1			
2		3		
0	↑		2	= $(231)_8$

Conversion from Decimal Fraction $(0.513)_{10}$ to Octal System

Example:

0.513 × 8 = 4.104
0.104 × 8 = 0.832
0.832 × 8 = 6.656

$0.656 \times 8 = 5.248$
$0.248 \times 8 = 1.984$
$0.984 \times 8 = 7.872$
$(0.513)_{10} = (0.406517...)_8$

The conversion of a decimal numbers with both the integers and fraction parts can be done by converting the integer and fraction separately and then the two answers shall be combined.

Conversion from binary to Hexadecimal

Example:

$$(10\ 1100\ 0110\ 1011.\ 1111\ 0000\ 0110)_2 = (2C6B.F06)_{16}$$

Conversion from Hexadecimal to binary

Example:

$$(306.D)_{16} = (0011\ 0000\ 0110.\ 1101)_2$$

Conversion from binary to Octal

Example:

$$(10\ 110\ 001\ 101\ 011.\ 111\ 100\ 000\ 110)_2 = (26153.7406)_8$$

Conversion from Octal to binary

Example:

$$(673.124)_8 = (110\ 111\ 011.\ 001\ 010\ 100)_2$$

Conversion from Hexadecimal to Decimal

Example:

$$(37B)_{16} = 3\times16^2 + 7\times16^1 + 11\times16^0$$

$$= 3\times256 + 7\times16 + 11\times1 = 768 + 112 + 11 = (891)_{10}$$

Complements

They are used to simplify the subtraction operation and it is also used for logical manipulation.

r-1's Compliments (Diminished Radix Complement)

Given a number N in base r which has n digits, then the (r – 1)'s complement of N is defined as (r – 1) – N. For decimal numbers, say r = 10 and r – 1 = 9, therefore the ninth complement of N is $(10^n - 1) - N$. Here, the 10^n represents a number that consists of a single 1 followed by n 0's. So the $10^n - 1$ is a number represented by n 9's.

Example:

- Take $n = 4 \rightarrow 10^4 = 10{,}000$ and $10^4 - 1 = 9999$.
- The 9's complement of 012398 is 999999 – 012398 = 987601.
- The 9's complement of 546700 is 999999 – 546700 = 453299.

For binary numbers, say r = 2 and r – 1 = 1, therefore the 1's complement of N is (2n – 1) – N. 2n is represented by a binary number that consists of 1 followed by n 0's.

Example:

- Take $n = 4 \rightarrow 2^4 = (10000)_2$ and $2^4 - 1 = (1111)_2$.
- The 1's complement of a binary number is formed by changing 0's to 1's and 1's to 0's.
- The 1's complement of 1011000 is 0100111.
- The 1's complement of 0101101 is 1010010.

r's Compliments (Radix Complement)

The r's complement of *n*-digit number N in base r is defined as $r^n - N$ for $N \neq 0$ and 0 for N = 0. By comparing with the (r – 1)'s complement, the r's complement is obtained by adding 1 to the (r – 1)'s complement since $r^n - N = \left[(r^n - 1) - N\right] + 1$.

Example:

- The 2's complement of binary $\rightarrow$ 101100 = $(2^6 - 1)$ – 101100 + 1 = 010100.
- The 10's complement of decimal $\rightarrow$ 2389 = $(10^4 - 1)$ – 2389 + 1 = 7611.
- The 10's complement of decimal $\rightarrow$ 246700 = $(10^6 - 1)$ – 246700 + 1 = 753300.
- The 10's complement of decimal $\rightarrow$ 012389 = $(10^6 - 1)$ – 012389 + 1 = 987602.

The 2's complement can be formed by leaving the least significant 0's and first 1 which is unchanged and replacing 1's with 0's and 0's with 1's in other four most-significant digits.

Example:

- The 2's complement of binary 1101100 is 0010100.

Example:

- The 2's complement of binary 0110111 is 1001001.

The radix complement and diminished radix complement are defined as:

$(N)_r$ = an n-digit number N in base r

$(N)_r$ = the r's complement of $(N)_r$

$(N)_{r-1}$ = the (r-1)'s complement of $(N)_r$

$$(N)_r = r^n - (N)_r \quad \text{...(1)}$$

$$(N)_{r-1} = [N]_r - 1 \quad \text{...(2)}$$

From the equation (1) and (2), we can also derive the following equations as follows:

$$[N]_r = [N]_{r-1} + 1 \quad \text{...(3)}$$

$$[N]_{r-1} = (r^n - 1) - (N)_r \quad \text{...(4)}$$

In the binary system, r = 2, we have 2's complement and 1's complement.

In the decimal system, r = 10, we have 10's complement and 9's complement.

In the octal system, r = 8, we have 8's complement and 7's complement.

Table: Represents the r's and the r-1's Compliments

System	Radix Complement (r's Compliments)	Diminished Radix Complement (r-1's Compliments)
Decimal	10's complement	9's complement
Octal	8's complement	7's complement
Binary	2's complement	1's complement

To find the radix complement representation of a number, first we shall derive the r-1's compliments (diminished radix complement) since it is more convenient. Then the r's compliments (radix complement) can be obtained by adding 1 to the r-1's compliments (diminished radix complement):

- The 1's complement of a binary integer is obtained by subtracting each digit from 1.
- The 7's complement of an octal number is obtained by subtracting each digit from 7.
- The 9's complement of a decimal number is obtained by subtracting each digit from 9.

Examples:

1. Let us find the 2's complement and the 1's complement of $(00011010)_2$:

   $(11100101)_2$ 1's complement

   $(11100110)_2$ 2's complement (add 1 to the 1's complement)

2. Let us find the 8's complement and the 7's complement of $(526071)_8$:

   $(251706)_8$ 7's complement

   $(251707)_8$ 8's complement (add 1 to the 7's complement)

3. Let us find the 10's complement and the 9's complement of $(546700)_{10}$:

   $(453299)_{10}$ 9's complement

   $(453300)_{10}$ 10's complement (add 1 to the 9's complement)

Signed Binary Numbers

It is customary to represent the sign with a bit placed in the leftmost position of the number and in order to make it 0 for positive and 1 for negative.

Let we consider the number 9 is represented in binary with 8 bits. Therefore the +9 is represented with sign bit 0 in the leftmost position which is followed by the binary equivalent of 9 to give 00001001.

Example:

00001001

Signed magnitude →10001001

Signed-1's complement → 11110110

Signed-2's complement → 11110111

Problems

Subtraction with Complements

1. By using 10's complement, Let us subtract 72532− 3250.

Solution:

Given:

M – N

M = 72532

10’s complement of N = + 96750 (99999 – 03250) + 1

Sum = 169282

Discard end carry 10^5 = - 100000

Answer =69282

2. Using 10’s complement, Let us subtract 3250– 72532.

Solution:

Given:

M – N

M = 03250

10’s complement of N = +27468 (99999 – 72532) + 1

Sum = 30718

No end carry

Answer : (10’s complement of 30718) = - 69282

3. Using 2’s complement, let us subtract 1010100 – 1000011.

Solution:

Given:

X – Y

X = 1010100

2’s complement of Y = + 0111101

Sum = 10010001

Discard end carry 2^7 = -10000

Answer: X - Y = 0010001

4. Using 2’s complement, let us subtract 1000011 – 1010100.

Solution:

Given:

Y – X

Y = 1000011

2's complement of X = +0101100

Sum = 1101111

No end carry.

Answer: Y – X - (2's complement of 1101111) = -0010001

5. Using 1's complement, let us subtract X – Y = 1010100 – 1000011.

Solution:

Given:

X = 1010100

1's complement of Y = +0111100 (+1 End-around carry)

Sum = 10010000

+ 1

Answer: X - Y = 0010001

6. Using 1's complement, let us subtract Y – X 1000011 – 1010100.

Solution:

Given:

Y = 1000011

1's complement of X = + 0101011

Sum= 1101110

No end carry

Answer: $Y - X$ - (1's complement of 1101110) = -0010001

1.2 4 Bit Codes, BCD, Excess-3, 2421, 84.2-1 9' Compliment Code and Other

The code and encoding is the process of assigning a group of binary bits to a discrete information, which can be present in the form of numbers, letters etc. Through assigning the item of information a unique combination of binary bits, we can transform that information to another form.

Therefore, these codes are used to communicate the information from or to a digital system. Generally the information is fed in the form of alphabets or some other symbols, the system then encodes this information and the processes by converting it by coding which is the processed information to the original form for the real world to interpret.

Thus, the encoding can be defined as the process of converting the symbolic information (digits, letters etc.) to special symbols such as binary. Therefore the transformed information is called as CODE. These bit combinations are called "WORDS" or "CODE-WORDS".

4 Bit Code

Base 10	Binary
0	0000
1	0001
2	0011
3	0010
4	0110
5	0111
6	0101
7	0100
8	1100
9	1101
10	1111
11	1110
12	1010
13	1011
14	1001
15	1000

Classification of Codes

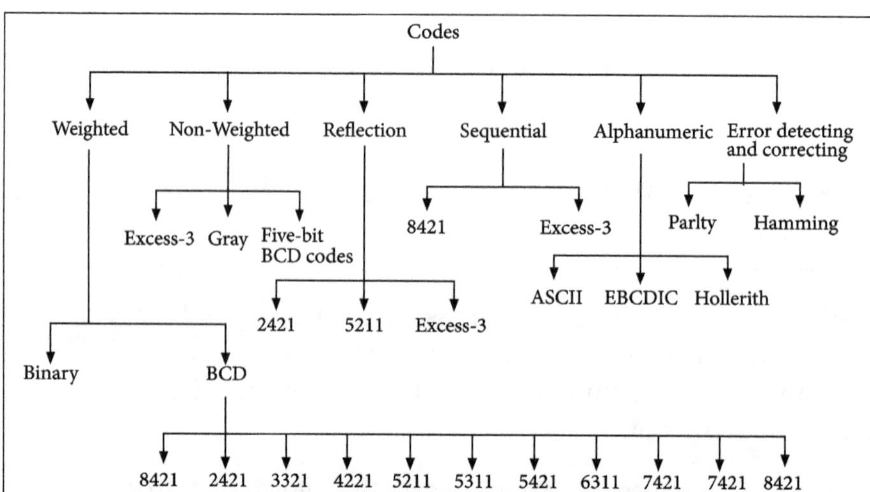

Classification of various binary codes.

There are many different coding schemes, where each has some advantages and characteristics. They are broadly classified into the following groups:

- Binary Codes:
    - Weighted Codes: BCD or 8421 Code, 2421.
    - Non-Weighted Codes: Excess-3 Code, Gray Code.
- Binary Coded Decimal Codes.
- Unit Distance Codes: Gray Code.
- Reflected Codes.
- Error Detecting and Correcting Codes.
- Alphanumeric Codes: ASCII code, EBCDIC.

Binary Codes

The binary code is the process of assigning the bit patterns to discrete elements of information. There can be different ways of assigning the bit patterns to this information, one way could be the straight forward assigning of binary codes which are binary equivalent of their numeric value. Another approach could be assigning some other combination of bits.

- Binary codes are codes which are represented in a binary system with modification from original ones.
- In general, N bits can represent up to 2^N distinct values.
- Conversely, to represent a range of M values, the number of bits required is $\log_2 M$.

Advantages of Binary Code

Following are the list of advantages that binary code offers:

- Binary codes are suitable for the computer applications.
- Binary codes are suitable for the digital communications.
- Binary codes make the analysis and designing of digital circuits if we use the binary codes.
- Since only 0 & 1 are being used, implementation becomes easy.

1 bit → represents up to 2 values (0, 1)

2 bit → represents up to 4 values (00, 01, 10, 11)

3 bit → represents up to 8 values (000, 001, 010, 011, 100, 101, 110, 111)

4 bit → represents up to 16 values (0000, 0001, 0010,..., 1110, 1111)

32 values → requires 5 bits

40 values → requires 6 bits

64 values → requires 6 bits

100 values → requires 7 bits

1024 values → requires 10 bits

- The base or radix of a number system is the number of digits present. The decimal numeral system has a base or radix of 10, where the set of 10 symbols (digits) is {0, 1, 2, 3, 4, 5, 6, 7, 8, 9}. The weights are in powers of ten.
- In general, a bases-b number $(a_n a_{n-1} \ldots a_o . f_1 f_2 f_m)_b$ has the value of,

$$\left(a_n \times b^n\right)+\left(a_{n-1} \times b^{n-1}\right)+\ldots+\left(a_o \times b^o\right)+\left(f_1 \times b^{-1}\right)+\left(f_2 \times b^{-2}\right)+\left(f_2 \times b^{-2}\right)+\ldots+\left(f_m \times b^{-m}\right)$$

Radix Number Systems

- A base of a number system or radix defines the range of values that a digit may have.
- In the binary system or base 2, there can be only two values for each digit of a number, either a "0" or a "1".
- In the octal system or base 8, there can be eight choices for each digit of a number: "0","1","2","3","4","5","6","7".
- In the decimal system or base 10, there are ten different values for each digit of a number: "0","1","2","3","4","5","6","7".
- In the hexadecimal system, we allow 16 values for each digit of a number: "0", "1", "2", "3", "4", "5", "6", "7", "8", "9", "A", "B", "C", "D", "E" and "F". Where "A" stands for 10, "B" for 11.

Table: Two Different ways of Assigning Binary Codes

Decimal code	Binary code format - 1	Binary code format - 2
0	0000	0000
1	0001	0010
2	0010	0100
3	0011	0110

4	0100	1000
5	0101	1001
6	0110	0111
7	0111	0101
8	1000	0011

The relation between the number of bits in code word and the number of unique code words is given by,

$$N = 2^j$$

Where,

- j is the number of bits.
- N is the total number of code words.

By taking log on both sides we get:

$$\text{Log}\,N = \log 2^j = j \log 2 >= j \text{i.e., } J = \log N \text{ (log N base 2)}$$

In log with base 10 form:

We get, j > = 3.32 log N (base 10)

For example:

5-bit codes can have 2^5 i.e., 32 total unique code, while the 6-bit code can have total of 64 code.

The binary code can be of following types:

- Weighted Binary Codes: Weighted binary codes are those which obey the positional weighting principles, each position of a number represents a specific weight. Example: Decimal, BCD, 8421, 2421, 5211, etc.
- Non-weighted binary code: Non-weighted codes are codes that are not positionally weighted. That is each position within the binary number is not assigned a fixed value. The decimal value of the code cannot be found by using the positional weight. Example: Gray code, Excess 3 code (XS-3), etc.

Excess-3 code is obtained by adding 3 to the decimal number. XS-3 is a self-complementing code, i.e., Complement of 0 (0011) is 9 (1100), complement of 1 (0100) is 8 (1011). It makes the code very useful for decimal subtraction.

Example:

$$\text{Decimal} \longrightarrow 8421\,\text{BCD} \xrightarrow[0011]{\text{Add}} \text{Excess}-3$$

Decimal	BCD 8 4 2 1	Excess-3 BCD + 0011
0	0 0 0 0	0 0 1 1
1	0 0 0 1	0 1 0 0
2	0 0 1 0	0 1 0 1
3	0 0 1 1	0 1 1 0
4	0 1 0 0	0 1 1 1
5	0 1 0 1	1 0 0 0
6	0 1 1 0	1 0 0 1
7	0 1 1 1	1 0 1 0
8	1 0 0 0	1 0 1 1
9	1 0 0 1	1 1 0 0

Let us consider an example to convert the code to its decimal equivalent or vice-versa.

Example: Convert decimal 543 to XS-3 code.

Decimal Code:	5	4	3
Add 3			
Result	8	7	6
XS-3 BCD Code	1000	0111	0110

Binary Coded Decimal Codes

It is simply said as BCD codes are the weighted binary codes. The other examples of weighted codes are 2421, 5421, 5411 code. These codes obey their positional weighting principles. The bits are multiplied by their weights; the sum of these weighted binary bits gives the equivalent decimal digit in the weighted binary system. All the BCD codes are used to represent the decimal numbers. Some other examples of BCD codes are, excess-3 Code (XS-3), 84-2-1 code (+8, +4, -2, -1), Biquinary code, 7421 code, 5311 code etc.

Table: Example of BCD codes

Decimal	XS-3	8421	84-2-1	Biquinary ABCDEFG	2421	5311	5421
0	0011	0000	0000	0100001	0000	0000	0000
1	0100	0001	0111	0100010	0001	0001	0001
2	0101	0010	0110	0100100	0010	0011	0010
3	0110	0011	0101	0101000	0011	0100	0011
4	0111	0100	0100	0110000	0100	0101	0100
5	1000	0101	1011	1000001	1000	0111	1000
6	1001	0110	1010	1000010	1001	1001	1001
7	1010	0111	1001	1000100	1010	1011	1010
8	1011	1000	1000	1001000	1011	1100	1011

The BCD codes have certain properties which make it very distinctive and thus have the advantage over the other code. These properties are: a. Self-complementing property b. Reflective property.

Self-complementing Codes

In the BCD system some code has self-complementing property that is their logical complement is the same as its arithmetic equivalent.

If the 9's complement of XS-3 is the same as its logical complement which helps to find the extensive use in decimal arithmetic.

Example:

631-1 code, 2421 code, XS-3 code.

We can verify the self-complementing property of these codes from the above table.

Reflective Codes: It is the second property, which makes the BCD code simply change in one bit, this code is reflected about the center entries with one bit changed. A code is said to be reflective when the code for 9 is a complement for the code for 0 and so for 8 and 1 code, 7 and 2, 6 and 3, 5 and 4. Codes 2421, 5211 and excess-3 are reflective, whereas the 8421 code is not.

It is said that the 9's complement of a reflected BCD code word is formed simply by changing only one of its bit.

Example:

2421 code, 5211 code and Excess (XS-3) code.

Sequential Code

A code is said to be sequential when two subsequent codes seen as numbers in the binary representation, differ by one. Each successive code is one binary number greater than the preceding code. The 8421 and excess-3 codes are sequential, whereas 2421 and 5211 codes are not.

Examples:

8421, XS-3.

BCD Code

In computing and electronic systems, binary-coded decimal (BCD) is an encoding for decimal numbers in which each digit is represented by its own binary sequence:

- Its main virtue is that it allows easy conversion to decimal digits for printing or display and faster decimal calculations.

- In BCD, a digit is usually represented by four bits which, in general, represent the values/digits/characters 0-9.

To BCD-encode a decimal number using the common encoding, each decimal digit is stored in a four-bit nibble. The position weights of the BCD code are 8, 4, 2, 1.

Decimal	0	1	2	3	4	5	6	7	8	9
BCD	0000	0001	0010	0011	0100	0101	0111	0111	1000	1001

- Thus, the BCD encoding for the number 127 would be: 0001 0010 0111

(0001 0010 0111) → BCD equivalent of 127

Whereas the pure binary number would be $(01111111)_2$

Example:

```
              ADD (148 + 157 ) = ?

   148   --BCD-->     0001        0100        1000
   157   --BCD-->     0001        0101        0111
                    -----------------------------------
                      0010        1001        1111
                         1 <-.    0110
                    ---------'             -----------
                      0011      ----1 <----  10101
                                  10000
                    -----------------------------------
Answer                   3          0           5
```

When the sum of 2 digits is greater than or equal to 9, then we need to add 6 i.e., 0110.

BCD to Excess - 3 code encoder

Excess - 3 codes is a modified form of a BCD number. It can be derived from the natural BCD code by adding 3 to each coded number.

Truth Table:

B3	B2	B1	B0	E3	E2	E1	E0
0	0	0	0	0	0	1	1
0	0	0	1	0	1	0	0
0	0	1	0	0	1	0	1
0	0	1	1	0	1	1	0
0	1	0	0	0	1	1	1
0	1	0	1	1	0	0	0
0	1	1	0	1	0	0	1
0	1	1	1	1	0	1	0
1	0	0	0	1	0	1	1

Advantages of BCD Codes

- It is similar to the decimal system.
- The binary equivalent of decimal numbers 0 to 9 only.

Disadvantages of BCD Codes

- The BCD arithmetic is little more complicated.
- The addition and subtraction of BCD have different rules.
- BCD needs more number of bits than the binary for representing the decimal number. So, the BCD is less efficient than binary.

Weighted Code Number Systems

Decimal Digit	8 4 2 1 Code	8 4 - 2 -1 code	2 4 2 1 code
0	0 0 0 0	0 0 0 0	0 0 0 0
1	0 0 0 1	0 1 1 1	0 0 0 1
2	0 0 1 0	0 1 1 0	0 0 1 0
3	0 0 1 1	0 1 0 1	0 0 1 1
4	0 1 0 0	0 1 0 0	0 1 0 0
5	0 1 0 1	1 0 1 1	1 0 1 1
6	0 1 1 0	1 0 1 0	1 1 0 0
7	0 1 1 1	1 0 0 1	1 1 0 1
8	1 0 0 0	1 0 0 0	1 1 1 0
9	1 0 0 1	1 1 1 1	1 1 1 1

Gray Code: This code is unweighted code. The gray code exhibits only a single bit change from one code number to the next. This is a variable weighted code and is cyclic.

Binary to Gray Code Conversion

1. Write down the number in binary codes.

2. The Most Significant Bit (MSB) of the gray code will be same as the MSB of binary code.

3. Perform XOR operation on MSB and then to the next bit to the MSB in binary number.

4. Repeat step 3 till all bits of binary number have been XO Red, the resultant code is the gray code equivalent to binary code.

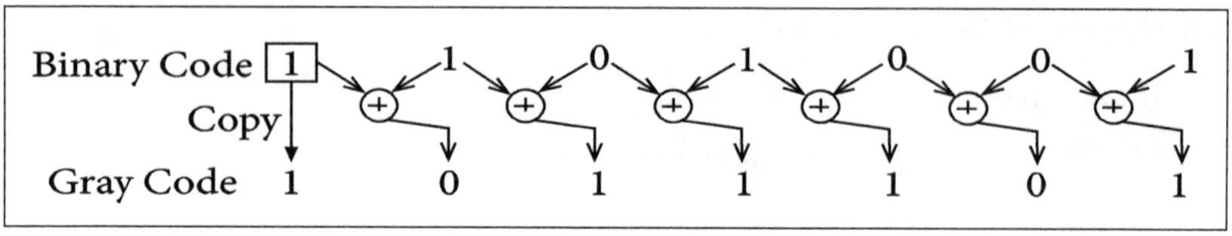

Conversion of binary code to gray code.

- MSB in the gray code is same as corresponding digit in binary number.
- Starting from left to right, add each adjacent pair of binary digits to get next and gray code digit (discard the carry if generated).

Gray to Binary Conversion

1. Start with the MSB of gray coded number.

2. Copy this bit as the MSB of the binary number.

3. Now, perform Ex-OR operation of this bit with the next bit of the binary number.

4. Repeat step 3 till all bits of gray coded number have been used in XOR operation. The resultant number is the binary equivalent of the gray number.

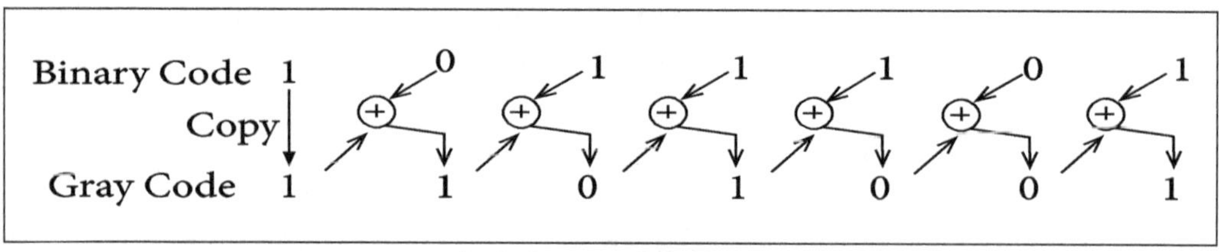

Conversion of gray code to binary code.

- MSB of binary is same as that of gray code.
- Add each binary digit to the generated gray digit in the next adjacent position (discard the carry if generated).

Excess-3 Code: The excess-3 is a non-weighted code used to represent decimal numbers. It can be derived by adding 3 to each decimal digit and then converting the result to four bits binary.

The code derives its name from the fact that each binary code is the corresponding 8421 code plus 0011 (3).

Example:

Decimal	8421	Excess-3
8	1000	1000 + 0011 = 1011
6	110	0110 + 0011 =1001

2421 Code

This is a weighted code; its weights are 2, 4, 2 and 1. A decimal number is represented in 4 bit form and the total 4 bits weight is 2 + 4 + 2 + 1 = 9. Hence, 2421 code represents the decimal numbers from 0 to 9.

Decimal	0	1	2	3	4	5	6	7	8	9
2421	0	1	10	11	100	1011	1100	1101	1110	1111

Complements: Complements are used in digital computer system for simplifying the subtraction operation and for logical manipulation. There are two types of complements for each base r system:

1. The r's Complement: Given a positive number N with base r with an integer part of n digits. This complement of N is defined as $r^n - N$ for $N \neq 0$ and 0 for N = 0. e.g., 10's complement of (25.639)10 is (102 - 25.639) 100 - 25.639 = 74 - 361, here the number of digits in integer part is 2 means n = 2

2. The (r -1)'s Complement: Given, a positive number N in base r with an integer part of n digits and a fraction part of m digits, the (r - 1)'s complement of N is defined as $r^n - r^{-m} - N$.

- 1's complement of $(52520)_{10}$ is $\left(10^5 - 1 - 52520\right) = 99999 - 52520 = 47479$
  Because number of integer part is 5, so $r^n = 105$ and no fractional part is Present' so $r^{-m} = 10^{-0} = 1$.
- 1's complement of $(0.3267)_{10}$ is $\left(10^0 - 10^{-4} - 0.3267\right) = 1 - 0.0001 - 0.3267$ $= 0.9999 - 0.3267 = 0.6732$ No integer part, so $10^n = 10^0 = 1$.
- 1's complement of $(101100)_2$ is
  $\left(2^6 - 2^0\right)10 - (101100)2 = (64 - 1)_{10} - (101100)_2 = (63)_{10}$
  $= (101100)_2 = 111111 - 101100 = 010011$

Key Points:

- 520 -f Here, n = 3, but (052) $\rightarrow$ here n = 2.
- In the latter example, 0 is of no significance.

Representation of Integers

There are three possible ways to represent a number. They can be given as:

- Signed magnitude method
- 1's complement method

- 2's complement method

1. Signed Magnitude Method: In this method, Number is divided into two parts, one is sign bit and other part for magnitude. In example we are using 5 bit register to represent -6 and +6.

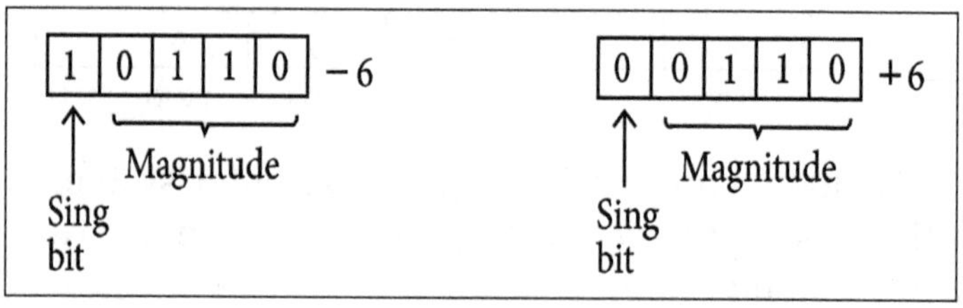

Range of Number: For n bit register, MSB will be a sign bit and (n - 1) bits will be magnitude.

- Drawback of signed magnitude method is that 0 will be having 2 different representation one will be 10000 i.e., -0 and the other one will be 00000 + 0.

2. 1's Complement Method: In this method, Positive numbers are represented in same way as in sign magnitude method. If number is negative, then it is represented using 1's complement method. For this, we first need to represent the number with positive sign and then take 1's complement of this number.

Example:

Suppose we are using 5 bit register. The representation of -6 will be represented as below:

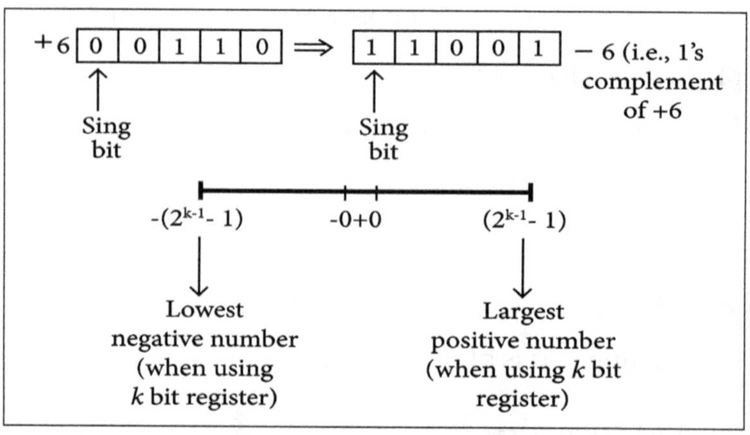

1's Complement method.

Drawback of 1's complement method is that there are two different representation for zero, one is -0 and other is +0.

3. 2's Complement Method: In 2's complement method, Positive numbers are

represented in same way as in sign magnitude. For representing negative number, we take 2's complement of the corresponding positive number.

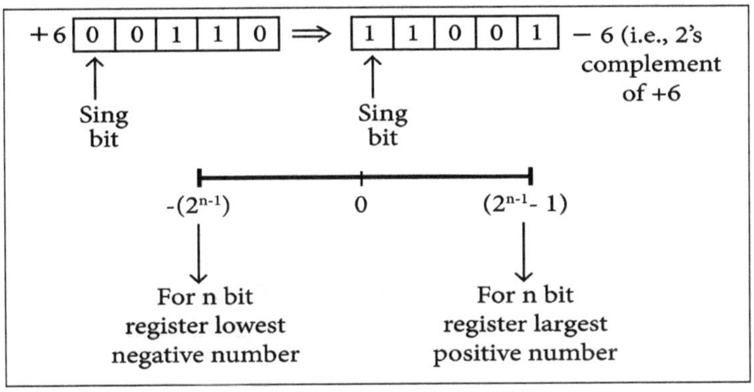

2's Complement method.

- 2's complement representation allows the use of binary arithmetic operations on signed integers, yielding the current 2's complement result.
- Positive Numbers: Positive 2's complement numbers are represented as the simple binary.
- Negative Numbers: Negative 2's complement numbers are represented as the binary number that when added to a positive number of the same magnitude equals zero.

Morse Code

This is the code, where the numbers, alphabets are represented as dots and dashes and this representation is known as Code.

Use of Encoding

Encoding is used to standardize a set of universal codes that can be used by all. The encoding is used when data security is of importance.

Alphanumeric codes | ASCII code | EBCDIC Code | UNICODE

Alphanumeric codes are sometimes called character codes due to their certain properties. Now these codes are basically binary codes. We can write alphanumeric data, including data, letters of the alphabet, numbers, mathematical symbols and punctuation marks by this code which can be easily understandable and can be processed by the computers.

Input output devices such as keyboards, monitors; mouse can be interfaced using these codes. 12-bit Hollerith code is the better known and perhaps the first effective code in the days of evolving computers in early days.

During this period punch cards were used as the inputting and outputting data. But nowadays these codes are termed obsolete as many other modern codes have evolved. The most common alphanumeric codes used these days are ASCII code, EBCDIC code and Unicode.

ASCII Code

The full form of ASCII code is American Standard Code for Information Interchange. It is a seven bit code based on the English alphabet.

In 1967 this code was first published and since then it is being modified and updated. ASCII code has 128 characters some of which are enlisted below to get familiar with the code.

DEC	OCT	HEX	BIN	Symbol	HTML Number	Description
0	000	00	00000000	NUL	&#000;	Null char
1	001	01	00000001	SOH	&#001;	Start of Heading
2	002	02	00000010	STX	&#002;	Start of Text
3	003	03	00000011	ETX	&#003;	End of Text
4	004	04	00000100	EOT	&#004;	End of Transmission
5	005	05	00000101	ENQ	&#005;	Enquiry
6	006	06	00000110	ACK	&#006;	Acknowledgment
7	007	07	00000111	BEL	&#007;	Bell
8	010	08	00001000	BS	&#008;	Back Space
9	011	09	00001001	HT		Horizontal Tab
10	012	0A	00001010	LF
Line Feed
11	013	0B	00001011	VT	&#011;	Vertical Tab
12	014	0C	00001100	FF		Form Feed
13	015	0D	00001101	CR
Carriage Return
14	016	0E	00001110	SO	&#014;	Shift Out / X-On
15	017	0F	00001111	SI	&#015;	Shift In / X-Off

EBCDIC Code

The EBCDIC stands for Extended Binary Coded Decimal Interchange Code. IBM invented this code to extend the Binary Coded Decimal which existed at that time. All the IBM computers and peripherals use this code. It is an 8 bit code and therefore can accommodate 256 characters. Below is given some characters of EBCDIC code to get familiar with it:

Char	EBCDIC	HEX	Char	EBCDIC	HEX	Char	EBCDIC	HEX
A	1100 0001	C1	P	1101 0111	D7	4	1111 0100	F4
B	1100 0010	C2	Q	1101 1000	D8	5	1111 0101	F5
C	1100 0011	C3	R	1101 1001	D9	6	1111 0110	F6

D	1100 0100	C4	S	1110 0010	E2	7	1111 0111	F7
E	1100 0101	C5	T	1110 0011	E3	8	1111 1000	F8
F	1100 0110	C6	U	1110 0100	E4	9	1111 1001	F9
G	1100 0111	C7	V	1110 0101	E5	blank	...	...
H	1100 1000	C8	W	1110 0110	E6	.	...	...
I	1100 1001	C9	X	1110 0111	E7	(	...	...
J	1101 0001	D1	Y	1110 1000	E8	+	...	...
K	1101 0010	D2	Z	1110 1001	E9	$	...	...
L	1101 0011	D3	0	1111 0000	F0	*	...	...
M	1101 0100	D4	1	1111 0001	F1	)	...	...
N	1101 0101	D5	2	1111 0010	F2	-	...	...
O	1101 0110	D6	3	1111 0011	F3	/	...	...

1.3 Logic Operations, Error Detection and Correction Codes

Error Detecting Codes

When the digital information in the binary form is transmitted from one circuit or system to the circuit or system, an error may occur. To maintain the data integrity between transmitter and receiver, the extra bit or more than one bit are added in the data. The data along with the extra bit/bits forms the code.

Codes which allow only error detection are called error detecting codes. This means a signal corresponding to 0 may change to 1 or vice-versa due to presence of noise. To maintain the data integrity between transmitter and receiver, extra bit or more than one bit is added in the data. These extra bits allow the detection and some times, correction of error in the data.

Codes which allow only error detection are called error detecting codes and codes which allow error detection and correction are called error detecting and correcting codes.

The most common cause for errors are that the noise creep into the bit stream during the course of transmission from transmitter to the receiver. If these errors are not detected and corrected, the result could be worst as the digital systems are too much sensitive to errors and will malfunction due to the slightest of errors in transmitted codes.

The various methods of error detection and correction such as addition of extra bits that is called check bits, sometimes they are also called redundant bits as they don't have any information in them. The various codes used for error detection and correction code in digital system are:

- Two-dimensional Parity check
- Simple Parity check

- Checksum
- Cyclic redundancy check

Error Correcting Codes

Error Correction can be handled in two ways:

- Once when an error is discovered, the receiver might have the sender to retransmit the entire data unit. This is termed as backward error correction.
- The receiver can use an error-correcting code which automatically corrects certain errors. This is known as forward error correction.

In the theory, it is possible to correct any number of errors automatically. The error-correcting codes are more sophisticated than the error detecting codes and need more redundant bits. The number of bits required correcting the multiple-bit or burst error is so high that in most of the cases it is inefficient to do so. For this same reason, most error correction is limited to one, 2 or at the most three-bit errors.

Single-bit Error Correction

The concept of error-correction can be understood by examining the simplest case of single bit errors. A single bit error shall be detected by addition of a parity bit (VRC) with the data which required to be sent. A single additional bit can detect error, but it is not enough to correct that error too. For collecting an error one has to know the exact position of error, i.e., exactly which bit is in error?

For example, to correct the single-bit error in an ASCII character, the error correction must determine which one of the seven bits is in error. To do this we have to add some additional redundant bits. For calculating the numbers of redundant bits (r) required to correct (d) data bits, let us determine the relationship between the two. Therefore, we have (d + r) as the total number of bits which are to be transmitted, then r should be able to indicate at least d + r + 1 different values.

Of these, one value means no error and remaining d + r values indicate error location of the error in each of d+r locations. So, d + r + 1 state must be distinguishable by r bits and r bits can indicates 2^r states. Thus, 2^r must be greater than d+r+1, i.e., 2r > d + r + 1. The value of r must be determined by putting in the value of d in the relation.

Example:

If d is 7, the smallest value of r that satisfies the above relation is 4. Thus, the total bits which are to be transmitted is 11 bits (d+r =7+4 = 11).

1.3.1 Basic Logic Operations: NOT, OR, AND, Universal Building Blocks

Binary logic deals with variables that take on two discrete values and with operations that assume logical meaning. The two values the variables assume may be called by different names (true and false, yes and no, low and high, etc.), but for our purpose, it is convenient to think in terms of bits and assign the values 1 and 0. The binary logic introduced is equivalent to an algebra called Boolean algebra.

Definition of Binary Logic

The binary logic has of binary variables and a set of logical operations. The variables are designated by letters of the alphabet, such as A, B, C, X, y, z, etc.. with each variable having 2 and only two distinct possible values: 1 and 0. There are 3 basic logical operations: AND, OR and NOT.

AND Gate

This operation is shown by a dot or by the absence of an operator. For example, x.y = z or xy = z is read "x AND y is equal to z." The logical operation AND is interpreted to mean that z = 1 if and only if x = 1 and y = 1; otherwise z = 0. (Remember that x, y and z are binary variables and can be equal either to 1 or 0 and nothing else).

Y = A AND B AND C.......N

Y = A.B.C......N

Y = ABC......N

Logic diagram:

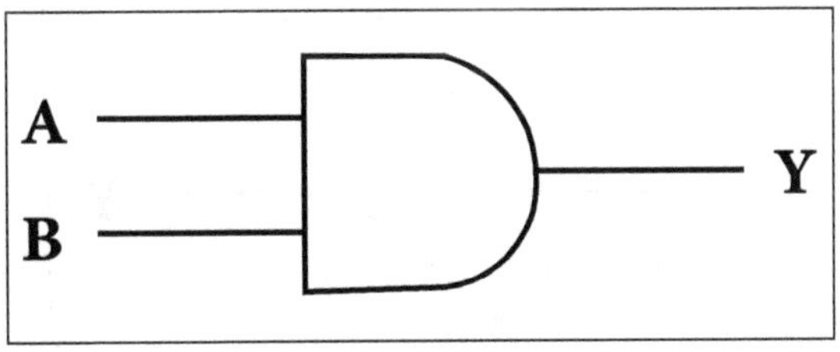

OR Gate

This operation is represented by a plus sign. For example, x + y = z is read "x OR y is equal to z" meaning that z = 1 if x = 1 or if y = 1 or if both x = 1 and y = 1. If both x = 0 and y = 0, then z = 0.

Y = A OR B OR C.......N

Y = A+B+C.......N

Logic diagram:

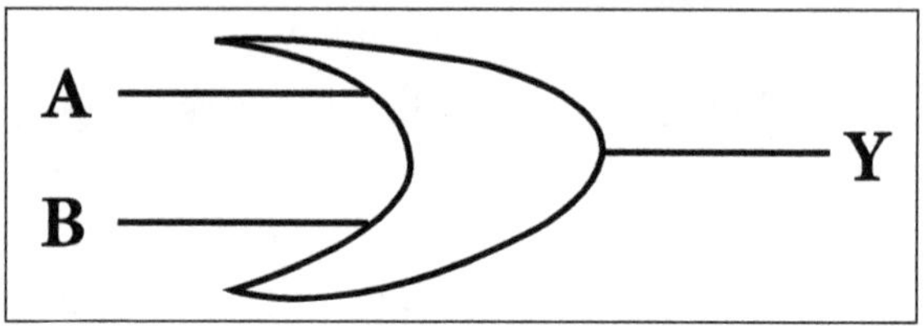

NOT

This operation is represented by a prime (sometimes by an over bar). For example, $x' = z$ (or $\bar{x}=z$) is read "not x is equal to z" meaning that z is what x is not. In other words. If x = 1. Then z = 0, but if x = 0, then z = 1. The NOT operation is also defined to as the complement operation, since it modifies a I to 0 and 0 to 1.

Binary logic resembles binary arithmetic and the operations AND and OR have similarities to multiplication and addition, respectively. In fact, the symbols utilized for AND and OR are the same as those used for multiplication and addition. However, the binary logic should not be confused with binary arithmetic. One should realize that an arithmetic variable designates a number that may consist of many digits. A logic variable is always either 1 or 0. For example, in binary arithmetic. We have 1 + 1 = 10 (read "one plus one is equal to 2"). Whereas in binary logic, we have 1 + 1 = 1 (read "one OR one is equal to one").

Y = NOT A

$Y = \bar{A}$

Logic diagram:

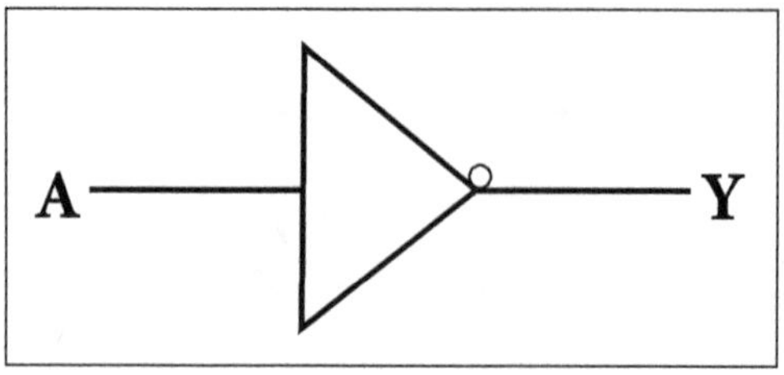

For every combination of the values of x and y, there is a value of z specified by the definition of the logical operation. Definitions of logical operations might be listed in a compact form called truth tables. A truth table is a table of all possible combinations of the variables, showing the relation between the values that the variables might take and the result of the operation.

The truth tables for the operations AND and OR with variables x and y are obtained by listing all possible values that the variables may have when combined in pairs. For each combination, the result of the operation is then listed in a separate row. The truth tables for AND, OR and NOT are given in the table below. These tables clearly demonstrate the definition of the operations.

Truth Tables of Logical Operations:

AND			OR			NOT	
x	y	x - y	x	Y	x + y	x	x’
0	0	0	0	0	0	0	1
0	1	0	0	1	1	1	0
1	0	0	1	0	0		
1	1	1	1	1	1		

Universal building blocks

NAND Gate

A NOT-AND operation is known as NAND operation. This gate has n input (n >= 2) and one output.

Y = A NOT AND B NOT AND C......N

Y = $\bar{A}$ NAND B NAND C.......N

Logic diagram:

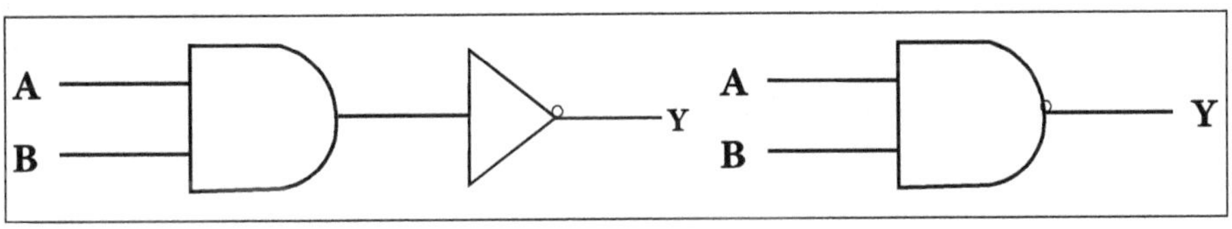

Truth Table

Inputs		Output
A	B	$\overline{AB}$
0	0	1
0	1	1
1	0	1
1	1	0

NOR Gate

The NOT-OR operation is known as NOR operation. This gate has n input $(n >= 2)$ and one output.

Y = A NOT OR B NOT OR C……N

Y = A NOR B NOR C…….N

Logic diagram:

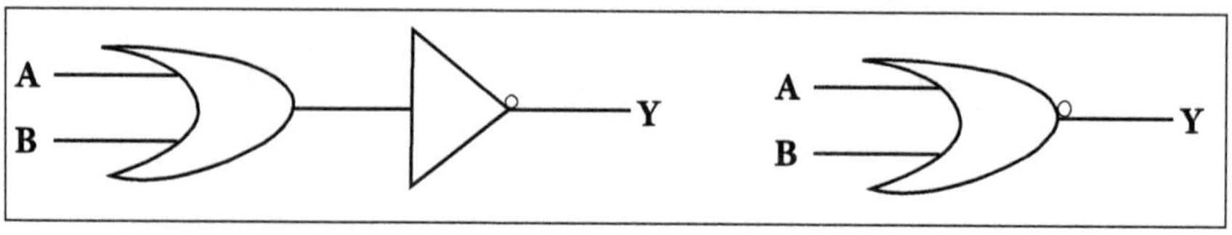

Truth Table

Inputs		Output
A	B	$\overline{A+B}$
0	0	1
0	1	0
1	0	0
1	1	0

EX-OR, EX-NOR - Gates

XOR Gate

The XOR or Ex-OR gate is a special type of gate. This gate can be used in the half adder, full adder and the subtractor. Here, the exclusive-OR gate can be abbreviated as EX-OR gate or sometime as X-OR gate. This gate has n input (n > = 2) and one output.

Logic diagram:

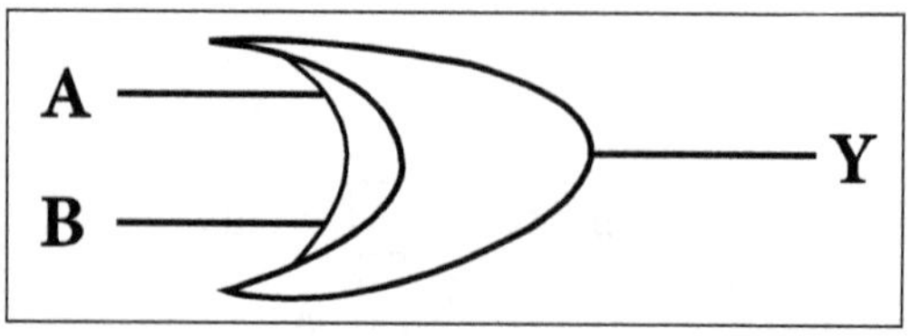

Truth Table

Inputs		Output
A	B	$A \oplus B$
0	0	0
0	1	1
1	0	1
1	1	0

XNOR Gate

The XNOR gate is a special type of gate. It can be used in the half adder, full adder and subtractor. The exclusive-NOR gate is shortly known as EX-NOR gate or sometime as X-NOR gate. It has n input (n > = 2) and one output.

$$Y = A \text{ XOR } B \text{ XOR } C......N$$

$$Y = A \ominus B \ominus C......N$$

$$Y = \bar{A}\bar{B} + AB$$

Logic diagram:

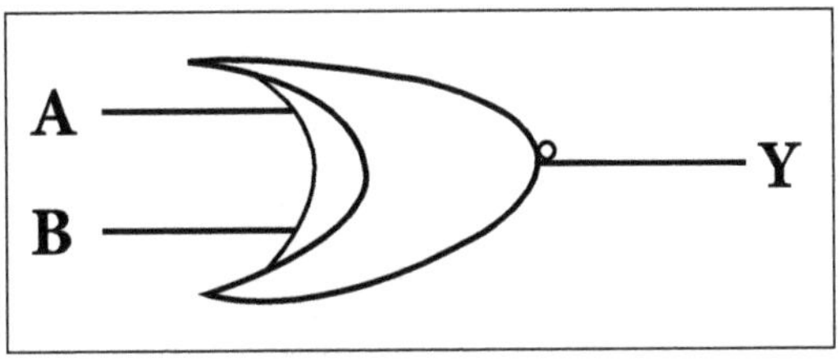

Truth Table

Inputs		Output
A	B	$A \ominus B$
0	0	0
0	1	1
1	0	1
1	1	0

1.3.2 Standard SOP and POS, Forms and Gray Code

Sum of Products (SOP)

A Sum-of-Products Boolean expression is literally a set of Boolean terms added together each term being a multiplicative combination of Boolean variables.

Sum-of-products-expression = term1 + term2 ... + term n

The product terms that include all the input variables are called as minterms.

In a sum-of-products expression, we form a product of all the input variables (or their inverses) for each row of the truth table for which the result is logic 1. The output is the logical "sum" of these minterms.

The Sum-of-Products expressions are easy to generate by determining which rows of the table have an output of 1, writing the one product term for each row and then finally sum all the product terms. This creates a Boolean expression which is representing the truth table as a whole.

The Sum-of-Products expressions lend themselves well to implementation as a set of AND gates (products) feeding into a single OR gate (sum).

Truth Table

A	B	C	W
0	0	0	0
0	0	1	1
0	1	0	1
0	1	1	1
1	0	0	0
1	0	1	1
1	1	0	1
1	1	1	0

The output can be expressed as,

$$W = \bar{A} \bullet \bar{B} \bullet C + \bar{A} \bullet B \bullet \bar{C} + \bar{A} \bullet \bar{B} \bullet C + A \bullet B \bullet \bar{C}$$

Product of Sums (POS)

An alternative to a Sum-of-Products expression to account for all "high" (1) output conditions in the truth table is to generate the Product-of-Sums or POS expression to account for all "low" (0) output conditions instead.

POS Boolean expressions may be generated from the truth tables quite easily, by determining the rows of the table which have an output of 0, writing the one sum term for each row and then finally multiply all the sum terms. This creates a Boolean expression which is representing the truth table as a whole.

These "sum" terms that include all the input variables (or their inverses) are called maxterms. In the POS implementation, the output variable is the logical product of maxterms. Product-Of-Sums expressions lend themselves well for implementation as a set of OR gates (sums) feeding into a single AND gate (product).

Truth Table

A	B	C	Z
0	0	0	1

0	0	1	1
0	1	0	1
0	1	1	1
1	0	0	0
1	0	1	0
1	1	0	0
1	1	1	1

The output can be expressed as,

$$Z = \left(\bar{A} + B + C\right) \bullet \left(\bar{A} + B + \bar{C}\right) + \left(\bar{A} + \bar{B} + C\right)$$

Problems

1. Given, a function $F = A + \bar{B}C$ expressed as a sum of its minterms is equivalent to a function expressed as a product of its maximum terms.

Solution:

Given:

$$F = A + \bar{B}C$$

Sum of Product (SOP):

$$F = A + \bar{B}C$$
$$= A\left(B + \bar{B}\right)\left(C + \bar{C}\right) + \bar{B}C\left(A + \bar{A}\right)$$
$$= \left(AB + A\bar{B}\right)\left(C + \bar{C}\right) + A\bar{B}C + \bar{A}\bar{B}C$$
$$F = ABC + A\bar{B}C + AB\bar{C} + A\bar{B}\bar{C} + A\bar{B}C + \bar{A}\bar{B}C$$
$$= ABC + AB\bar{C} + A\bar{B}C + A\bar{B}\bar{C} + \bar{A}\bar{B}C$$
$$= m_7 + m_6 + m_5 + m_4 + m_1$$
$$F = \Sigma\left(1, 4, 5, 6, 7\right)$$

2. Let us express the function $F = A + \bar{B}C$ in Canonical SOP form.

Solution:

Given:

$$F = A + \bar{B}C$$

Sum of Product (SOP):

$$Y = A+\overline{B},C = A\left(B+\overline{B}\right)+\overline{B}C\left(A+\overline{A}\right) = AB+A\overline{B}+A\overline{B}C+\overline{A}\overline{B}C$$
$$= AB\left(C+\overline{C}\right)+A\overline{B}\left(C+\overline{C}\right)+A\overline{B}C+\overline{A}\overline{B}\,C$$
$$= ABC+AB\overline{C}+A\overline{B}C+A\overline{B}\overline{C}+A\overline{B}C+\overline{A}\overline{B}C$$

$$Y = m_7 + m_6 + m_5 + m_3 + m_5 + m_1$$

$$Y = \Sigma(1, 4, 5, 6, 7)$$

3. Let us to express the function $F = A+\overline{B}C$ in Canonical POS form.

Solution:

Given:

$$F = A+\overline{B}C$$

Product of Sum (POS):

$$Y = A+\overline{B}C = \left(A+\overline{B}\right)\left(A+C\right) = [\therefore A+BC = (A+B)+(A+C)]$$
$$= \left(A+\overline{B}+C\right)+\left(A+\overline{B}+\overline{C}\right)+\left(A+B+C\right)+\left(A+\overline{B}+C\right)$$
$$= \left(A+\overline{B}+C\right).\left(A+\overline{B}+\overline{C}\right).\left(A+B+C\right)$$

$$Y = M_2 M_3 M_0 = M_0 M_2 M_3$$

$$Y = \pi\left(0,\ 2,\ 3\right)$$

4. Let us minimize and implement the following multiple output functions in SOP form.

$$f_1 = \Sigma m\ (0,2,6,10,11,12,13) + d\ (3,4,5,14,15)$$

$$f_2 = \Sigma m\ (1,2,6,7,8,13,14,15) + d\ (3,5,12)$$

Solution:

Given:

$$f_1 = \Sigma m\ (0,2,6,10,11,12,13) + d\ (3,4,5,14,15)$$

$$f_2 = \Sigma m\ (1,2,6,7,8,13,14,15) + d\ (3,5,12)$$

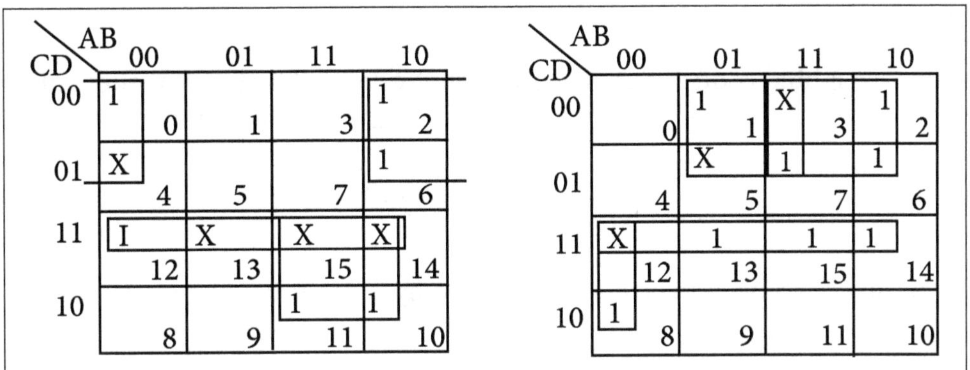

$f_1=\bar{B}\bar{C}+CD+AC$ $\quad$ $f_2=B\bar{C}+A\bar{C}+CD+\bar{A}\bar{B}C$

The K maps are filled with ones and don't cares using the expression. After reduction we find that CD occurs both in f_1 and f_2. So it can be shared.

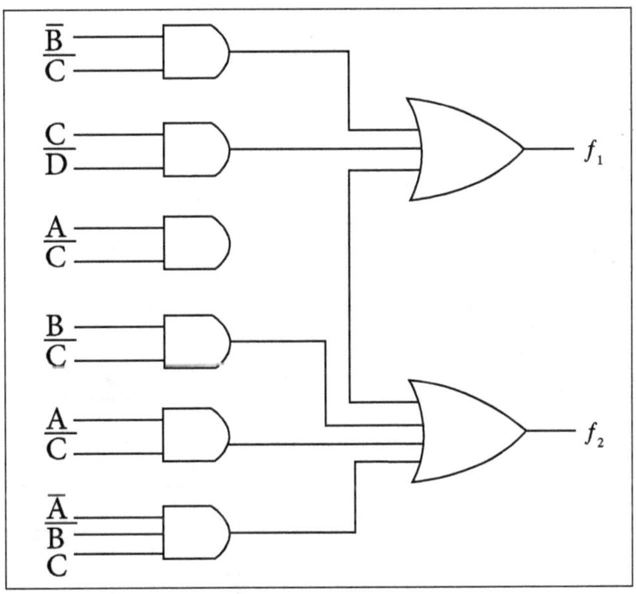

Logic Circuit Implementation.

5. Let us convert the given expression in canonical SOP form Y = AC + AB + BC.

Solution:

Given:

$$Y = AC + AB + BC$$
$$= AC\,(B + B') + AB\,(C + C') + (A + A')\,BC$$
$$= ABC + ABC' + AB'C + AB'C' + ABC + ABC' + ABC$$
$$= ABC + ABC' + AB'C + AB'C' \;[A + A = 1]$$

Gray Code

It is the non-weighted code and it is not arithmetic codes. That means there are no specific weights assigned to the bit position. It has a very special feature that, only one bit will change each time the decimal number is incremented. As only one bit changes at a time, the gray code is called as a unit distance code. The gray code is a cyclic code. Gray code cannot be used for arithmetic operation.

A binary number can be converted to gray code by using following steps:

- Record the MSB as it is.
- Add this bit to the next position, recording the sum and neglecting any carry.
- Record successive sums until completed.

Conversion of BCD to Gray Code

The truth table for BCD to gray code converter is:

BCD Code				Gray Code			
B3	B2	B1	B0	G3	G2	G1	G0
0	0	0	0	0	0	0	0
0	0	0	1	0	0	0	1
0	0	1	0	0	0	1	1
0	0	1	1	0	0	1	0
0	1	0	0	0	1	1	0
0	1	0	1	0	1	1	1
0	1	1	0	0	1	0	1
0	1	1	1	0	1	0	0
1	0	0	0	1	1	0	0
1	0	0	1	1	1	0	1

Applications of Gray Code

- Gray code is popularly used in the shaft position encoders.
- A shaft position encoder produces a code word which represents the angular position of the shaft.
- Gray code is used in applications where the normal sequence of binary numbers may produce an error or ambiguity during transition from one number to next.
- It is used in telegraphy, robust communication, error detection and correction.

K-map Simplification

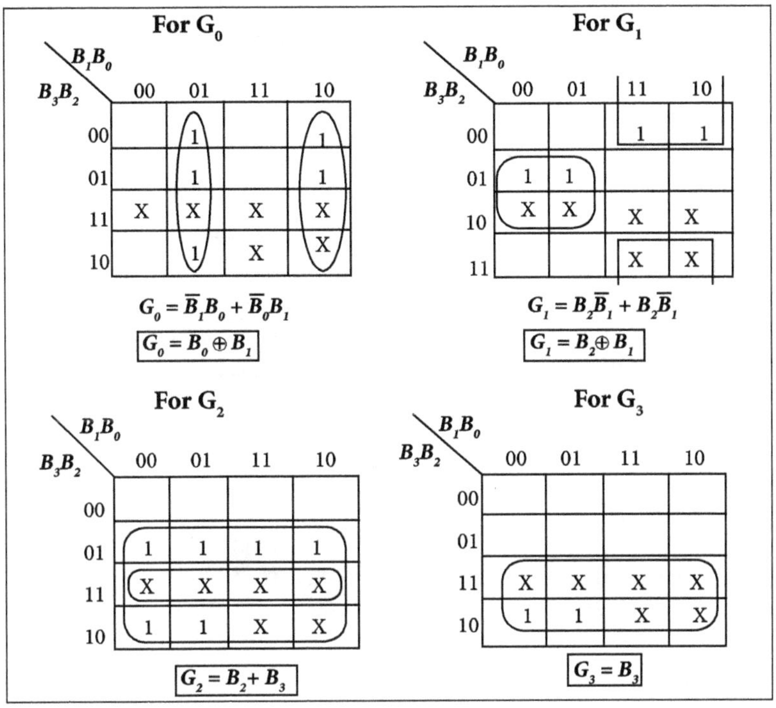

Logic Diagram:

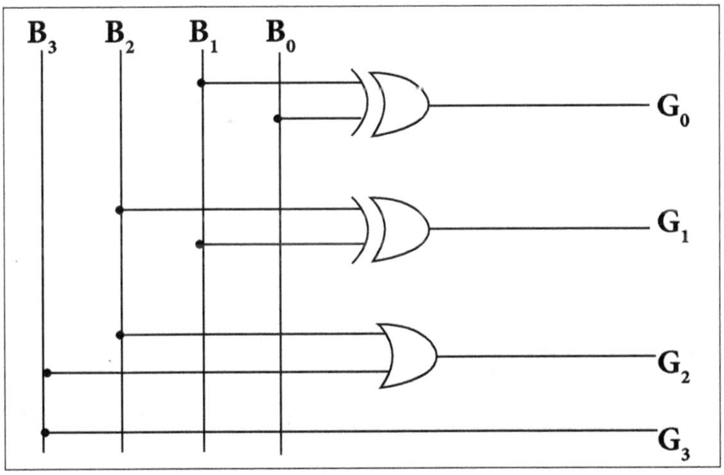

Conversion of Combinational Circuits from Gray Code to BCD.

Gray				BCD				
G3	G2	G1	G0	B4	B3	B2	B1	B0
0	0	0	0	0	0	0	0	0
0	0	0	1	0	0	0	0	1

0	0	1	0	0	0	0	1	0
0	0	1	1	0	0	0	1	1
0	1	0	0	0	0	1	0	0
0	1	0	1	0	0	1	0	1
0	1	1	0	0	0	1	1	0
0	1	1	1	0	0	1	1	1
1	0	0	0	0	1	0	0	0
1	0	0	1	0	1	0	0	1
1	0	1	0	1	0	0	0	0
1	0	1	1	1	0	0	0	1
1	1	0	0	1	0	0	1	0
1	1	0	1	1	0	0	1	1
1	1	1	0	1	0	1	0	0
1	1	1	1	1	0	1	0	1

1.3.3 Error Detection, Error Correction Codes (Parity Checking, Even Parity, Odd Parity And Hamming Code)

Parity Checking

Error Detection Codes

Basic approach used for error detection is the use of redundancy, where additional bits are added to facilitate detection and correction of errors. Popular techniques are as follows:

- Simple Parity check
- Two-dimensional Parity check
- Checksum
- Cyclic redundancy check

Simple Parity Checking or One-dimension Parity Check

The most common and least expensive mechanism for error- detection is the simple parity check. In this technique, a redundant bit called parity bit, is appended to every data unit so that the number of 1s in the unit (including the parity becomes even).

Blocks of data from the source are subjected to a check bit or Parity bit generator form, where a parity of 1 is added to the block if it contains an odd number of 1's (ON bits) and 0 is added if it contains an even number of 1's.

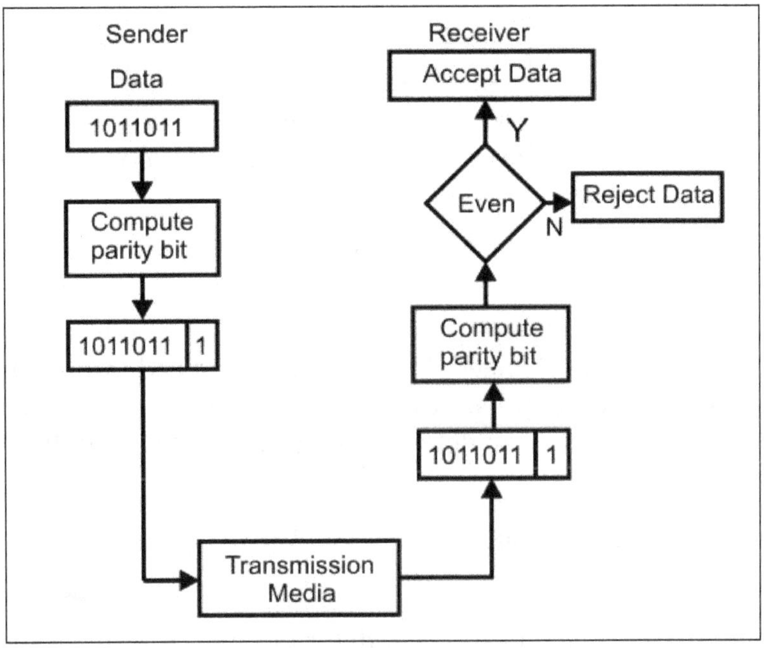

Even-parity checking scheme.

At the receiving end the parity bit is computed from the received data bits and compared with the received parity bit, as shown in the below figure. This scheme makes the total number of 1's even, that is why it is called even parity checking. Considering a 4-bit word, different combinations of the data words and the corresponding code words are given in below table.

Table: Possible 4-bit data words and corresponding code words.

Decimal value	Data Block	Parity bit	Code word
0	0000	0	00000
1	0001	1	00011
2	0010	1	00101
3	0011	0	00110
4	0100	1	01001
5	0101	0	01010
6	0110	0	01100
7	0111	1	01111
8	1000	1	1001
9	1001	0	10010
10	1010	0	10100
11	1011	1	10111
12	1100	0	11000
13	1101	1	11011

14	1110	1	11101
15	1111	0	11110

Note that for the sake of simplicity, we are discussing here the even-parity checking, where the number of 1's should be an even number. It is also possible to use odd-parity checking, where the number of 1's should be odd.

Performance

An observation of the table reveals that to move from one code word to another, at least two data bits should be changed. Hence these set of code words are said to have a minimum distance (hamming distance) of 2, which means that a receiver that has knowledge of the code word set can detect all single bit errors in each code word.

However, if two errors occur in the code word, it becomes another valid member of the set and the decoder will see only another valid code word and know nothing of the error.

Thus, errors in more than one bit cannot be detected. In fact it can be shown that a single parity check code can detect only odd number of errors in a code word.

Two-dimension Parity Check

Performance can be improved by using two-dimensional parity check, which organizes the block of bits in the form of a table. Parity check bits are calculated for each row, which is equivalent to a simple parity check bit. Parity check bits are also calculated for all columns then both are sent along with the data. At the receiving end these are compared with the parity bits calculated on the received data. This is given in below figure.

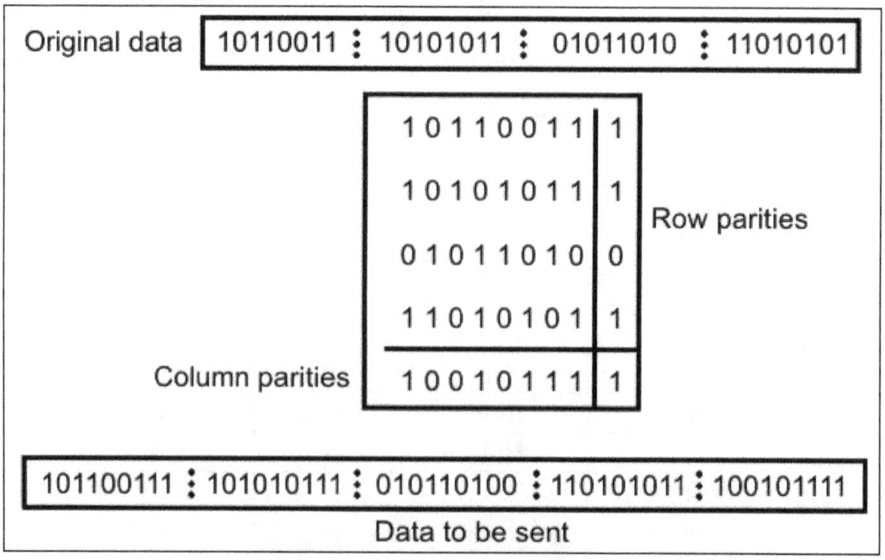

Two-dimension Parity Checking.

Performance

Two- Dimension Parity Checking increases the likelihood of detecting burst errors. As we have shown in the above figure that a 2-D Parity checks of n bits can detect a burst error of n bits. A burst error of more than n bits is also detected by 2-D Parity check with a high probability.

There is, however, one pattern of error that remains elusive. If two bits in one data unit are damaged and two bits in exactly same position in another data unit are also damaged, the 2-D Parity check checker will not detect an error. For example, if two data units: 11001100 and 10101100. If first and second from last bits in each of them is changed, making the data units as 01001110 and 00101110, the error cannot be detected by 2-D Parity check.

Parity Code

It is the simplest technique which is used for detecting and correcting errors. The MSB of an 8-bits word is used as the parity bit and the remaining 7 bits are used for data or message bits. Parity bit is added to the transmitted strings of bits during the transmission from the transmitters to detect any error in the data.

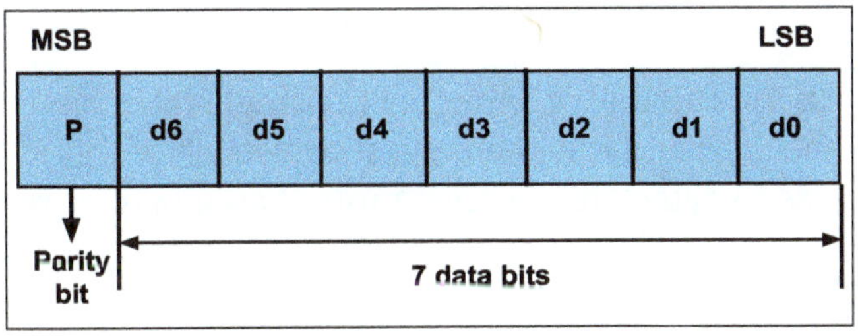

When they are received at the receiver end. It is an extra bit added to the string of data. The parity of 8-bits transmitted word can be either even parity or odd parity.

Parity checking at the receiver can detect the presence of an error if the parity of receiver signal is different from the expected parity. That means, if it is known that the parity of the transmitted signal is always even and if the received signal has an odd parity, then the receiver can conclude that the received signal is not correct. When an error is detected, then the receiver will ignore the received byte and request for the retransmission of the same byte to the transmitter.

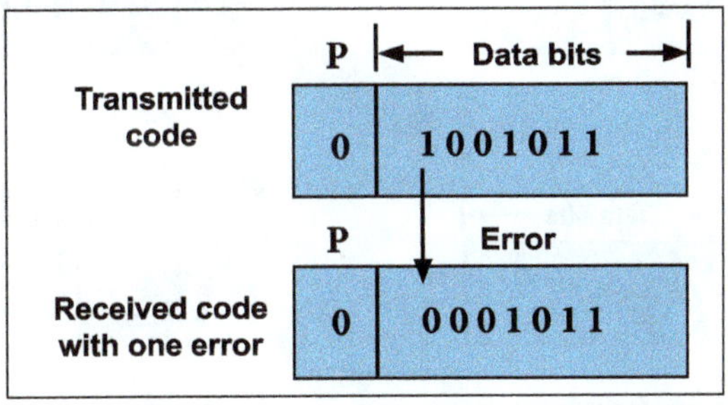

Even Parity

Even parity means the number of 1's in the given word including the parity bit should be even. When we get an even parity, after adding that extra bit the total numbers of 1's in the string of the data is even.

Odd Parity

Odd parity is defined as the number of 1's in the given word including the parity bit should be odd. When we get an odd parity, after adding that extra bit into the data string the total number of 1's in the data shall be odd.

We can understand it with an example, when we have an eight bit ASCII code – 01000001.

Now if the added bit is 0 then the number will become 001000001. In odd parity, the total number of 1's in the number is even so we get an even parity. Repeatedly if we add 1 to the number, then the number will become 101000001.

The number of 1's is 3 which is odd so we have got an odd parity. Then, normally even parity is used and it has almost become a convention.

Now the parity checks are capable of detecting the single bit error. Since, it fails if there are two changes in the data and it is the biggest drawback of this system. Hence, there are several other codes to detect and correct more than one bit errors.

Use of Parity Bit

The parity bit can be set to a 0 and 1 where it depends on the type of the parity required.

For even parity, this bit is set to a 1 or 0 whereas the no. of the 1 bits in the entire word is even which is shown in below figure (a).

For odd parity, this bit is set to 1 or 0 whereas the no. of the 1 bits in the entire word is odd which is shown in below figure (b).

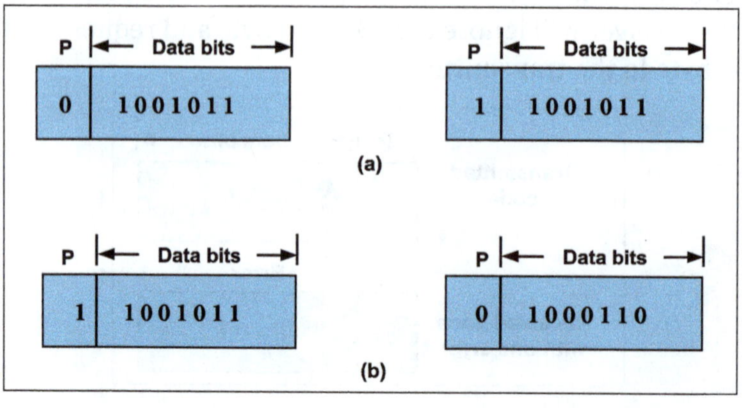

Hamming Code

Let us examine manipulation with these bits to discover which bit is in error. The technique developed by R.W. Hamming provides a practical solution. The solution or coding scheme, he developed is commonly called as Hamming Code.

The hamming code can be applied to data units of any length and which uses the relationship between the data bits and the redundant bits as discussed.

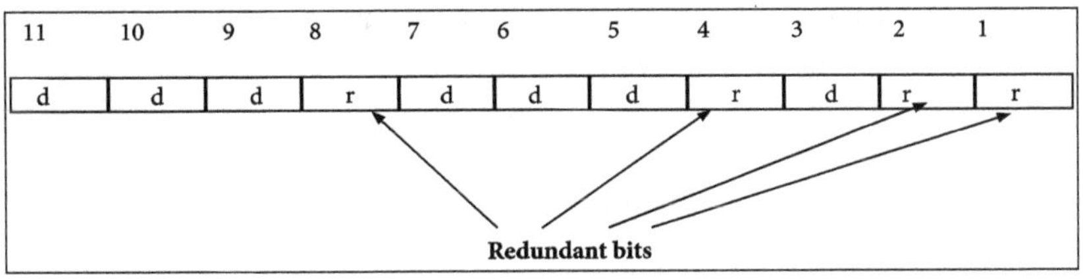

Positions of redundancy bits in hamming code.

The basic approach for error detection by using Hamming code is as follows:

- Location of each of the (m+K) digits is assigned a decimal value.
- To each group of m information bits k parity bits are added to form (m+k) bit code as shown in the above figure.
- At the receiving end, the parity bits can be recalculated. The decimal value of the k parity bits provides the bit-position in error, if any.
- The k parity bits are placed in positions 1, 2, 2K-1 and K parity checks are performed on selected digits of each code word.

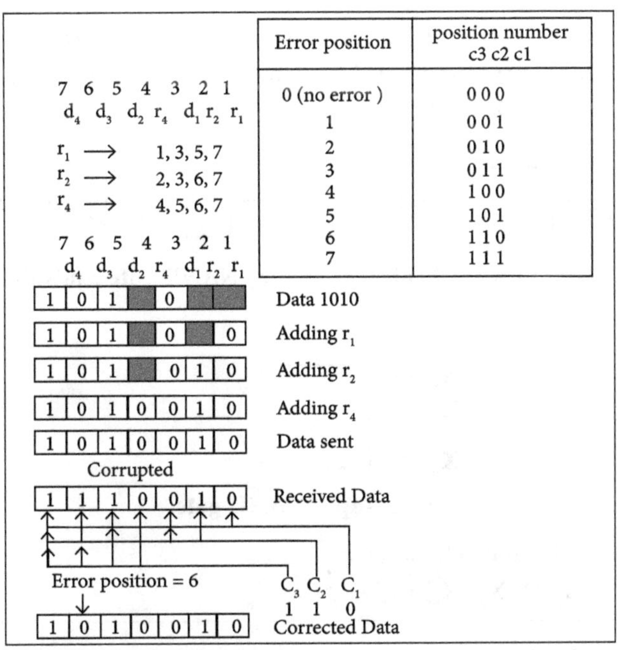

Error position	position number c3 c2 c1
0 (no error)	000
1	001
2	010
3	011
4	100
5	101
6	110
7	111

Use of Hamming code for error collection for a 4-bit data.

The above figure shows how the hamming code is used for the correction of 4-bit numbers ($d_4d_3d_2d_1$) with the help of three redundant bits $(r_3r_2r_1)$. For the example, consider data 1010. First r_1 (0) is calculated considering the parity of the bit positions 1, 3, 5 and 7. Then the parity bits r_2 is calculated by considering bit positions 2, 3, 6 and 7. Finally, the parity bits r_4 is calculated by considering the bit positions 4, 5, 6 and 7 as shown.

If any corruption occurs in any of the transmitted code 1010010, the bit position in the error can be found out by calculating the $r_3r_2r_1$ at the receiving end. For example, if the received code word is 1110010, the recalculated value of the $r_3r_2r_1$ is 110 that indicate that bit position in error is 6, the decimal value of 110.

Parity Generator: The circuit that generates the parity bit in the transmitter is called a parity generator.

Parity Checker: The circuit that checks the parity in the receiver is called parity checker.

1.4 NAND-NAND and NOR-NOR Realizations

The NAND and the NOR gates are generally faster and use fewer components than AND or OR gates:

- It can be implemented by using only NAND gates or only NOR gates with any logic function. NAND gate or NOR gate forms a functionally complete set where any switching function can be expressed in terms of NAND gate or NOR gate.
- Similarly, the set AND, OR and NOT shall form a functionally complete set since any Boolean function can be expressed in terms of the SOP and POS only by using the AND, OR and NOT operations.

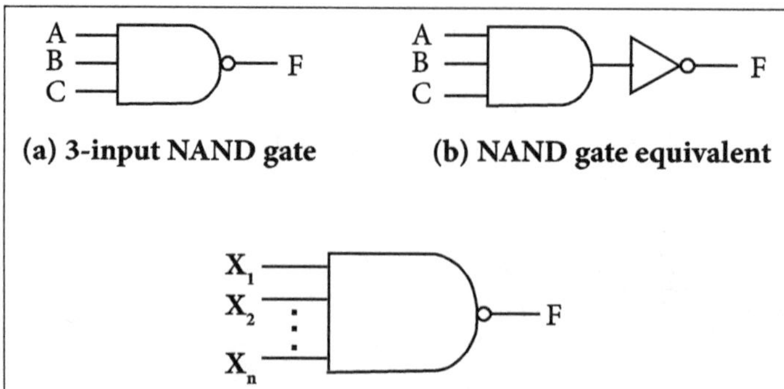

(a) 3-input NAND gate (b) NAND gate equivalent

(c) n-input NAND gate

N-input NAND gate:

$F=(X_1X_2...X_n)=X_1'+X_2'+...+X_n'$

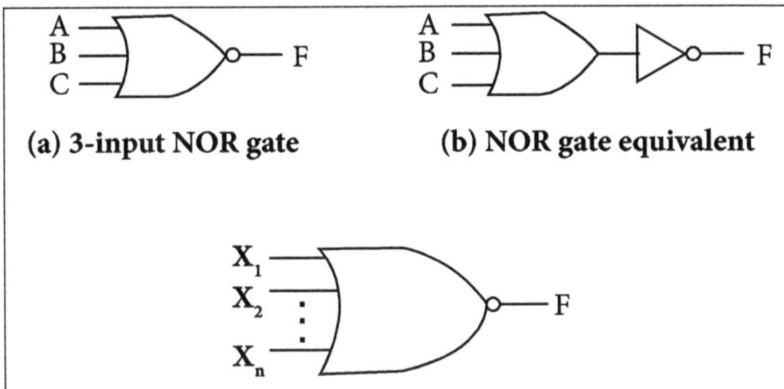

(a) 3-input NOR gate **(b) NOR gate equivalent**

(c) n-input NOR gate

N-input NAND gate:

$F=(X_1+X_2+...+X_n)=X_1'+X_2'...X_n'$

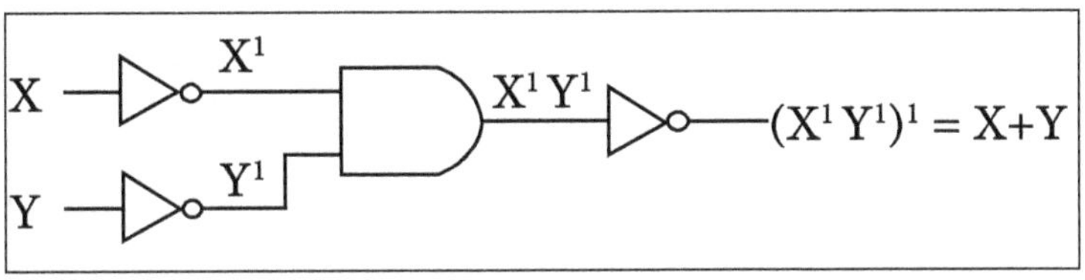

AND and NOT are a functionally complete set of gates.

OR can be realized using AND and NOT.

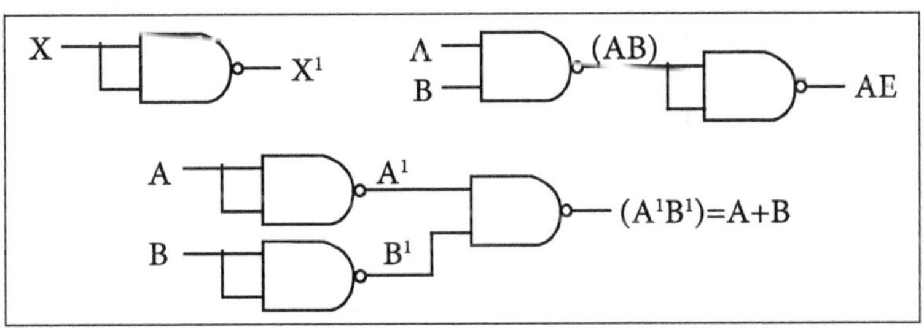

NAND Gate Realization of NOT, AND and OR

Design of 2-Level Circuits using NAND and NOR Gates.

Steps for Designing A Minimum 2-level NAND-NAND Circuit

- Find a minimum SOP expression for F.
- Then, draw the corresponding 2-level AND-OR circuit.
- Replace all gates with NAND gates leaving the gate interconnections unchanged. When the output gate has any single literals as inputs, complement these literals.

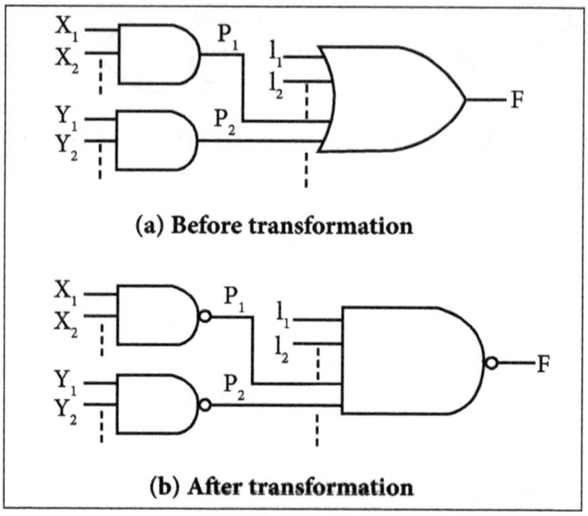

AND-OR to NAND-NAND Transformation.

Steps for Designing a Minimum 2-level NOR-NOR Circuit

- Find a minimum POS expression for F.
- Draw the corresponding 2-level OR-AND circuit.
- Then, replace all gates with NOR gates leaving the gate interconnections unchanged. When the output gate has any single literals as inputs and complements these literals.

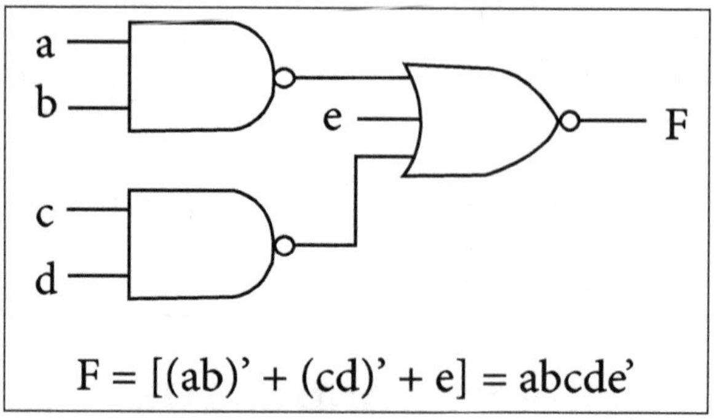

NOR-NOR Transformation.

Problems

1. Let us determine the NAND gate implementation.

$$F(x,y,z) = (1,2,3,4,5,7)$$

Solution:

Given:

$$F(x,y,z) = (1,2,3,4,5,7)$$

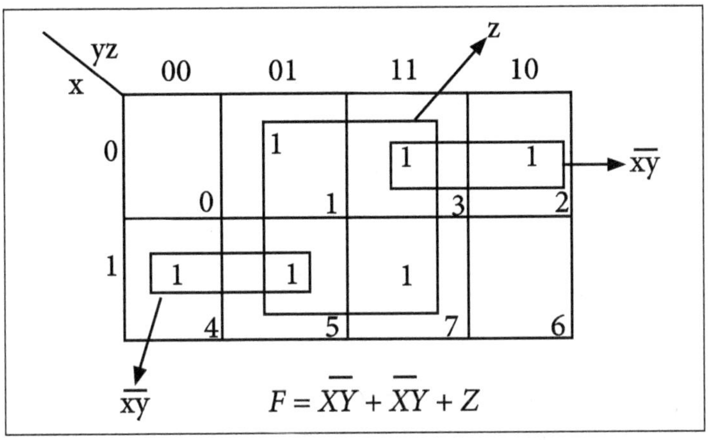

NAND Implementation:

(i)

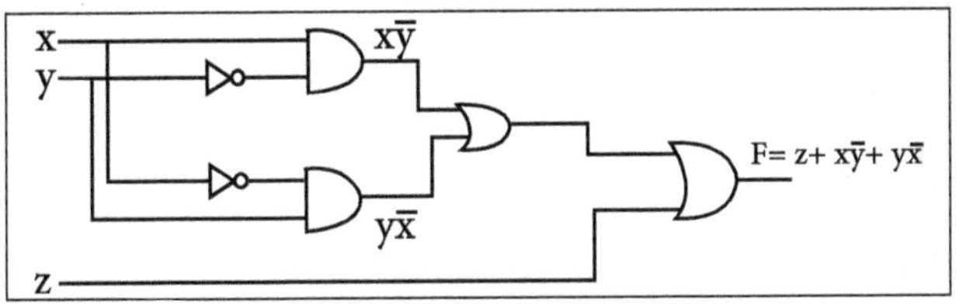

(ii)

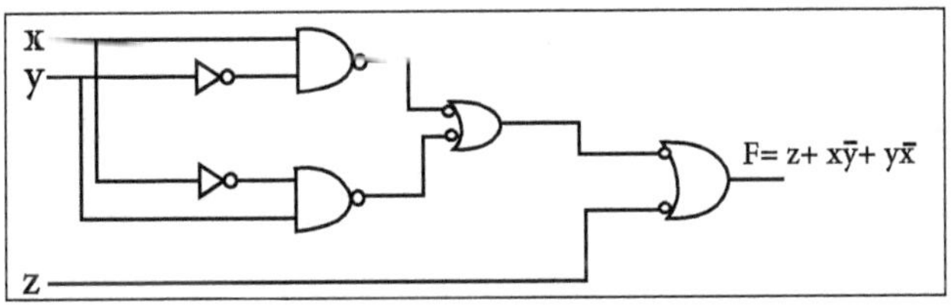

(iii)

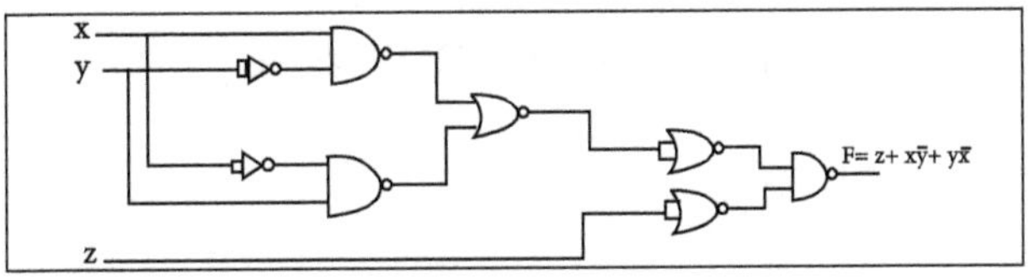

2. Let us implement a NAND logic diagram which has the complement of the functions F(A, B, C, D) = Σ (0, 1, 2, 3, 4, 8, 9, 12).

Solution:

Given:

$$F(A, B, C, D) = \Sigma(0, 1, 2, 3, 4, 8, 9, 12)$$

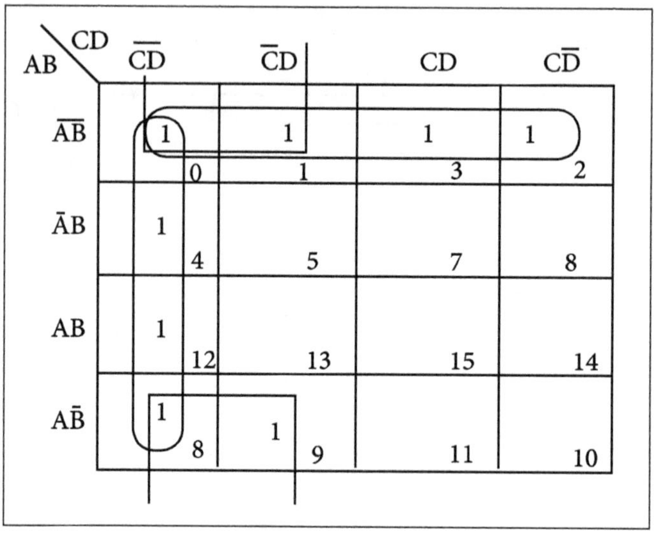

$$F = \bar{B}\bar{C}\bar{D} + \bar{B}\bar{C} + \bar{A}\bar{B}$$

$$F = \overline{\overline{\bar{C}\bar{D} + \bar{B}\bar{C} + \bar{A}\bar{B}}}$$

$$= \left(\overline{\bar{C}\bar{D}}\right)\left(\overline{\bar{B}\bar{C}}\right)\left(\overline{\bar{A}\bar{B}}\right)$$

$$= (C + D)(B + C)(A + B)$$

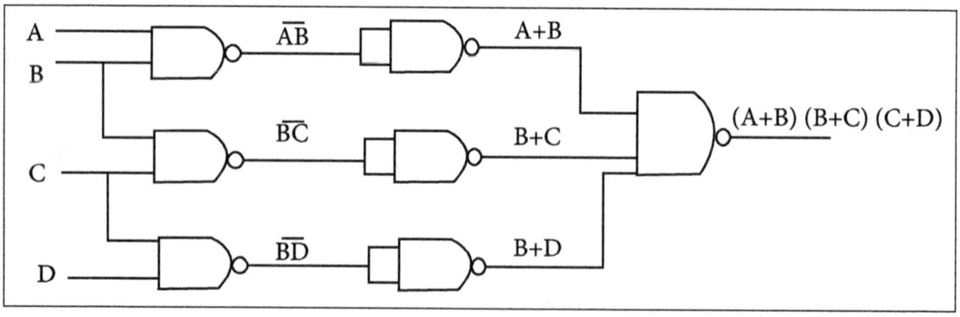

3. Let us implement EXOR using NAND and NOR gate.

Solution:

EXOR using NAND:

1)

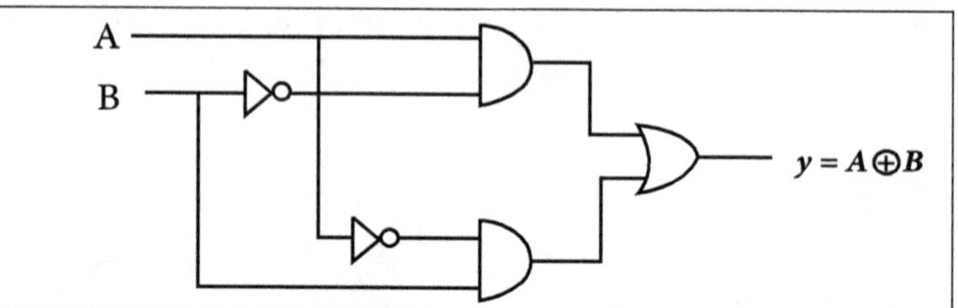

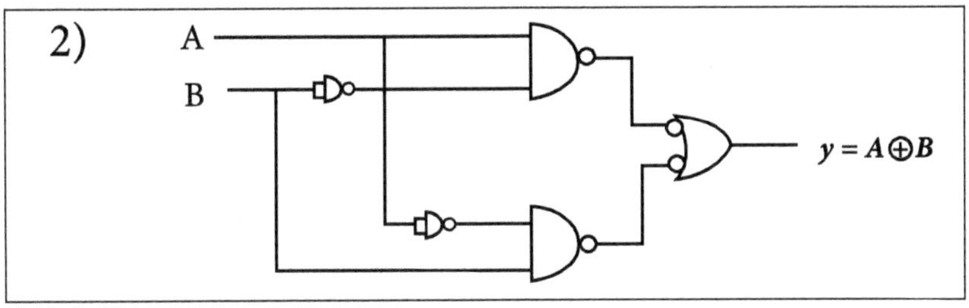

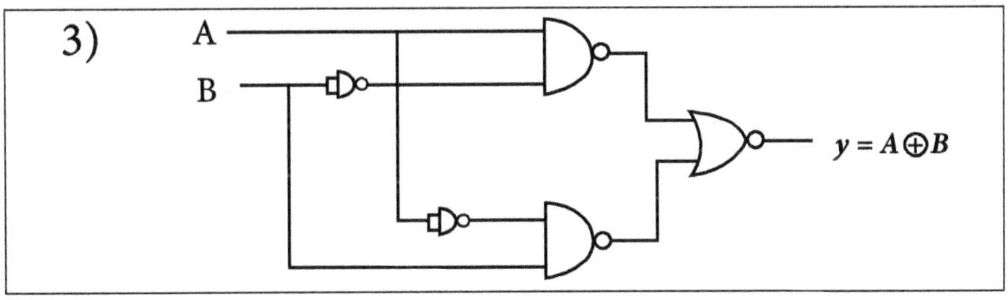

EXOR using NOR:

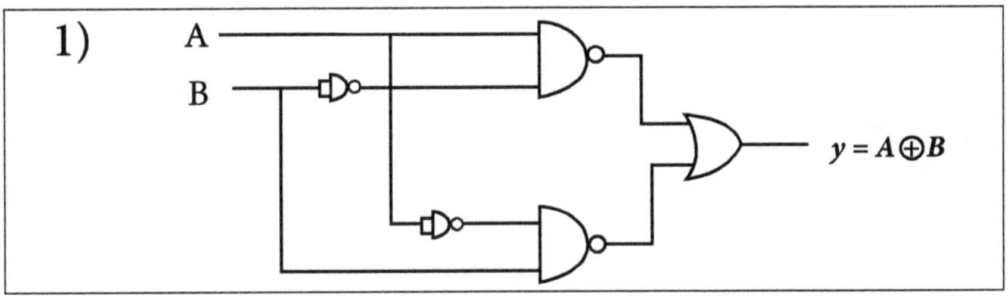

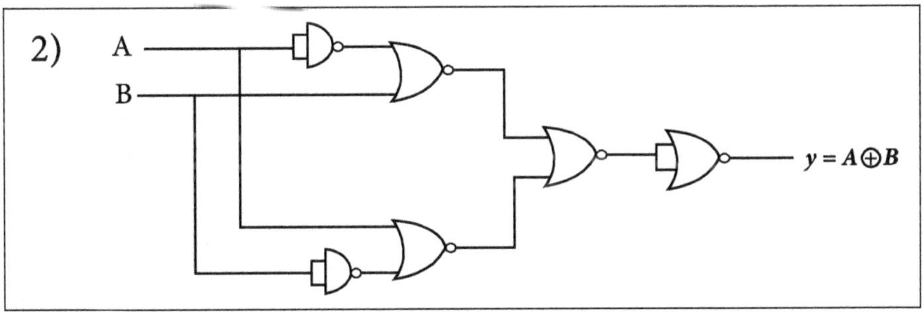

2

Minimization Techniques

2.1 Boolean Theorems, Principle of Complementation and Duality

Boolean Algebra is the mathematics which is used to analyze the digital gates and circuits. These “Laws of Boolean” can be used to reduce both and simplify the complex Boolean expression in an attempt to reduce the number of logic gates required. The Boolean Algebra is therefore, a system of mathematics which is based on logic that has its own set of rules or laws that are used to define and reduce Boolean expressions.

The variables which are used in Boolean Algebra only have one of two possible values, a logic “0” and a logic“1”, but an expression can have the infinite number of variables all labelled individually to represent the inputs to the expression. For example, the variables A, B, C etc., give us a logical expression of A + B = C, but each variable can only be either 0 or 1.

The examples for these individual laws of Boolean, rules and theorems of Boolean Algebra are given in the following table.

Truth Tables for the Laws of Boolean

Boolean Expression	Description	Equivalent Switching Circuit	Boolean Algebra Law or Rule
A + 0 = A	A in parallel with open = “A”	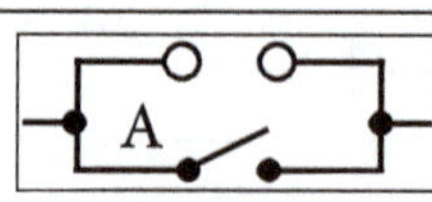	Identity
A + 1 = 1	A in parallel with closed = “CLOSED”	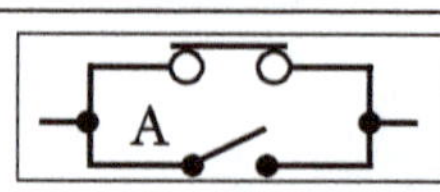	Annulment
A . 0 = 0	A in series with open = “OPEN”	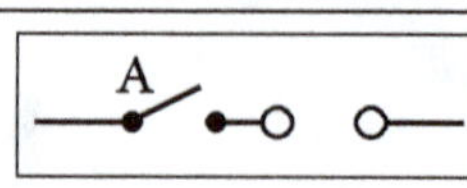	Annulment
A . 1 = A	A in series with closed = “A”	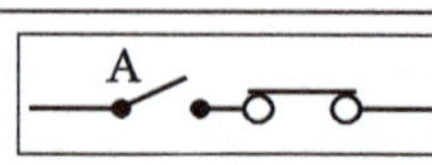	Identity
A . A = A	A in series with A = “A”	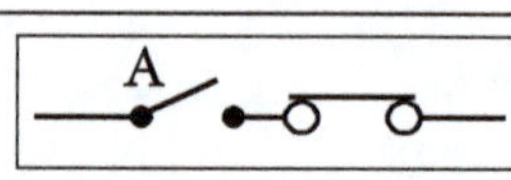	Idempotent

NOT $\bar{A} = A$	NOT NOT A(double negative) = "A"		Double Negation
$A + \bar{A} = A$	A in parallel with A = "A"	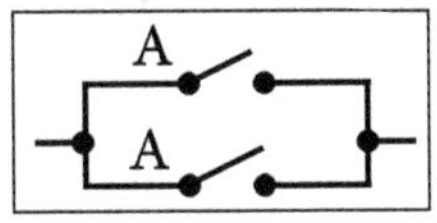	Idempotent
$A . \bar{A} = 0$	A in series with NOT A = "OPEN"	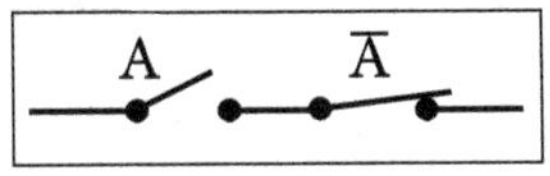	Complement
$A + A = 1$	A in parallel with NOT A = "CLOSED"	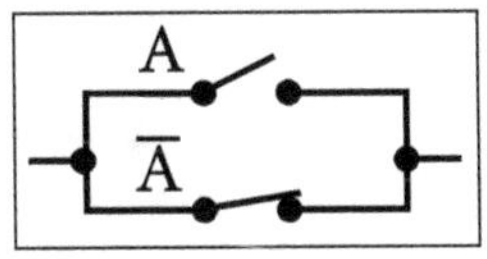	Complement
$A + B = B + A$	A in parallel with B =B in parallel with A	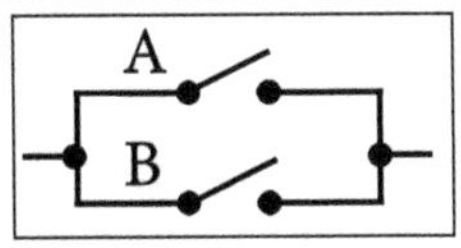	Commutative
$\overline{A+B} = \bar{A}\bar{B}$	Invert and replace OR with AND		De Morgan's Theorem
$A.B = B.A$	A in series with B =B in series with A	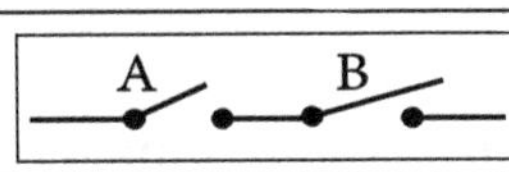	Commutative
$\overline{A+B} = \bar{A}\bar{B}$	Invert and replace AND with OR		De Morgan's Theorem

The basic laws of Boolean Algebra can relate to the Commutative Law allowing a change in position for the addition and multiplication, the Associative Law allowing the removal of brackets for the addition and multiplication, as well as the Distributive Law allowing the factoring of an expression, are the same as in ordinary algebra.

Every Boolean Laws given above is just a single or two variables, but the number of variables defined is by a single law which is not limited to this, as there can be an infinite number of variables as inputs to the expression. Those Boolean laws can be used to prove any given Boolean expression as well as for simplifying the complicated digital circuits.

Description of the Laws of Boolean Algebra

Identity Law

A term OR'ed with the value "0" or AND'ed with the value "1" will be always equal to that term.

Where,

- $A + 0 = A$A variable OR'ed with the value 0 is always equal to the variable.
- $1 = A$A variable AND'ed with the value 1 is always equal to the variable.

Annulment Law

The term AND'ed with a "0" equals 0 or OR'ed with a "1" will equal 1.

Where,

- 0 = 0A variable AND'ed with the value 0 is always equal to 0.
- A + 1 = 1A variable OR'ed with the value 1 is always equal to 1.

Idempotent Law

An input that is AND'ed or OR'ed with itself is equal to that input.

Where,

- A + A = AA variable OR'ed with itself is always equal to the variable.
- A . A = AA variable AND'ed with itself is always equal to the variable.

Complement Law

The term AND'ed with its complement equals to "0" and a term OR'ed with its complement equals to "1".

Where,

- $A.\bar{A} = 0$A variable AND'ed with its complement is always equal to 0.
- $A + \bar{A} = 1$A variable OR'ed with its complement is always equal to 1.

Commutative Law

An order of the application of two separate terms is not important.

Where,

- B = B. A The order in which two variables are AND'ed makes no difference.
- A + B = B + A The order in which two variables are OR'ed makes no difference.

Double Negation Law

The term that is inverted twice is equal to the original term.

Where,

- $\bar{A} = A$A double complement of a variable is always equal to the variable.
- The other algebraic Laws of Boolean includes.

Distributive Law

It permits the multiplying or factoring out of an expression.

Where,

- $A\left(B + C\right) = A.B + A.C$ OR Distributive Law).
- $A + \left(B.C\right) = \left(A + B\right).\left(A + C\right)$ (AND Distributive Law).

Absorptive Law

It enables the reduction in the complicated expression to a simpler one by absorbing like terms.

Where,

- A + (A.B) = A (OR Absorption Law).
- A(A + B) = A (AND Absorption Law).

Associative Law

It allows the removal of brackets from an expression and regrouping of the variables.

Where,

- A + (B + C) = (A + B) + C = A + B + C (OR Associate Law).
- $A\left(B.C\right)=\left(A.B\right)C=$ A. B. C (AND Associate Law).

Boolean Algebra Functions

By using the information above, a simple 2-input and OR and NOT gates could be represented by 16 possible functions as shown in the below following table.

Function	Description	Expression
1.	NULL	0
2.	IDENTITY	1
3.	Input A	A
4.	Input B	B
5.	NOT A	$\bar{A}$
6.	NOT B	$\bar{B}$
7.	A AND B (AND)	A.B
8.	A AND NOT B	$A.\bar{B}$
9.	NOT A AND B	$\bar{A}.B$

10.	NOT A AND NOT B (NAND)	$\overline{A.B}$
11.	A OR B (OR)	$A + B$
12.	A OR NOT B	$A+\overline{B}$
13.	NOT A OR B	$\bar{A} + B$
14.	NOT OR (NOR)	$\overline{A+B}$
15.	Exclusive-OR	$A.\overline{B}+\overline{A}.B$
16.	Exclusive-NOR	$A.B+\overline{A}.\overline{B}$

Example

Using the above laws, we simplify the following expression: (A + B) (A + C)

Q = (A + B).(A + C)

Q = A.A + A.C + A.B + B.C Distributive law

Q = A + A.C + A.B + B.C Idempotent AND law (A.A = A)

Q = A (1 + C) + A.B + B.C Distributive law

Q = A.1 + A.B + B.C Identity OR law (1 + C = 1)

Q = A (1 + B) + B.C Distributive law

Q = A.1 + B.C Identity OR law (1 + B = 1)

Q = A + (B.C) Identity AND law (A.1 = A)

Then the expression is given by:

(A + B)(A + C) can be simplified to A + (B.C) as in the Distributive law.

Duality Theorem

In mathematical optimization theory, duality or the duality principle is the principle in which optimization problems may be viewed from either of two perspectives, the primal problem or the dual problem. The solution to the dual problem provides a lower bound to the solution of the primal (minimization) problem.

The Duality Theorem starts with a Boolean relation. We can derive another Boolean relation by:

- Changing OR (operation) i.e., + (Plus) sign to an AND operation.
- Complement any 0 or 1 appearing in the expression.

Problems

1. Let us reduce the given expression using Boolean algebra.

$$x'y'z' + x'y'z' + x'yz + xy'z + xyz$$

Solution:

Given:

$$x'y'z' + x'y'z' + x'yz + xy'z + xyz = x'y'(z' + z) + yz(x' + x) + xy'z$$
$$= x'y' + yz + xy'z$$
$$= y'(x' + z) + yz$$
$$= y'x' + y'z + yz$$
$$= y'x' + z(y' + y)$$
$$= y'x' + z.$$

2. Let us reduce the given expression using Boolean algebra.

$$abc' + ab'c + a'bc + abc$$

Solution:

Given:

$$abc' + ab'c + a'bc + abc = ab(c' + c) + ab'c + a'bc$$
$$= ab + ab'c + a'bc = a(b + b'c) + a'bc$$
$$= ab + ac + a'bc = b(a + c) + ac = ab + ac + bc$$

3. Let us reduce the given expression using Boolean algebra.

$$p'q'r + p'qr' + p'qr + pqr' + pq'r'$$

Solution:

Given:

$$p'q'r + p'qr' + p'qr + pqr' + pq'r' = p'r(q' + q) + qr'(p' + p) + pq'r'$$
$$= p'r + qr' + pq'r'$$
$$= r'(q + q'p) + p'r$$
$$= qr' + pr' + p'r$$
$$= qr' + p \oplus r.$$

2.2 De-Morgan Theorems: Minimization of Logic Functions using Boolean Theorems

De-Morgan was a great logician and mathematician. Among his contribution, the following two theorems are important.

De-Morgan's First Theorem

De-Morgan's First Theorem states that "the complement of the sum of the variables is equal to the product of the complement of each variable".

It can be expressed by the following Boolean expression.

$$\overline{A+B} = \bar{A}.\bar{B}$$

A	B	A + B	$\overline{A+B}$ L.H.S	$\bar{A}$	$\bar{B}$	$\bar{A}.\bar{B}$ R.H.S
0	0	0	1	1	1	1
1	0	1	0	1	1	0
0	1	1	0	1	0	0
1	1	1	0	0	0	0

De-Morgan's Second Theorem

De-Morgan's Second Theorem states that the "Complement of the product of variables is equal to the sum of complements of each individual variables".

The Boolean expression for this theorem is known as Boolean Function.

$$\overline{A.B} = \bar{A}+\bar{B}$$

A	B	A.B	$\overline{A.B}$ L.H.S	$\bar{A}$	$\bar{B}$	$\bar{A}+\bar{B}$ R.H.S
0	0	0	1	1	1	1
1	0	0	1	0	1	1
0	1	1	1	1	0	1
1	1	1	0	0	0	0

Canonical Form of Boolean Expressions

It is the expanded form of Boolean expression, where each term contains all Boolean variables in their true or complemented form, is called as the canonical form of the expression.

De Morgan suggested two theorems that form important part of Boolean algebra. They are:

- The complement of a sum term is equal to the product of the complements.

  $(A + B)' = A'B'$

- The complement of a product is equal to the sum of the complements.

  $(AB)' = A' + B'$

2.2.1 Minimization of Logic Functions Using Boolean Theorems

The minimization technique is a process or a simplification procedure that are used to minimize the number of literals in a Boolean logic expression. The main objective of all the minimization techniques is to obtain an expression that has the minimum number of terms.

If we have more than one possible solution with the same number of terms, the one having the minimum number of literals is the choice. The process is usually called as logic minimization and the goal is to form a result which is efficient.

There are several methods for simplification of Boolean logic expressions. The most commonly used techniques are:

- Boolean algebra postulates and theorems,
- Karnaugh maps,
- Petrick's method,
- Quine-McCluskey method,
- Generalized concensus algorithm.

For very complicated problems, the Algebraic Minimization method can be done using special software analysis programs. K-maps are also limited to problems with up to 4 binary inputs. For more than 4 binary inputs, the Quine - McCluskey tabular method is used.

Boolean Postulates

The algebraic system usually used to work with binary logic expressions.

Postulates

- Closure: Any defined operation on (0, 1) gives (0,1)
- Identity: $0 + x = x + 0$; $1.x = x$

- Commutative: $x + y = y + x$; $x.y = y.x$
- Distributive: $x\left(y + z\right) = xy + xz$; $x + \left(yz\right) = \left(x + y\right)\left(x + z\right)$

For every element x B, there exists an element x' B (complement of x) such that:

- $x + x' = 1$
- $x.x' = 0$

There exists at least two elements x, y B such that $x = y$.

The precedence rule is based on the following term orders:

- parentheses
- NOT
- AND
- OR

Problems

1. Let us reduce the expression $\overline{A + \overline{BC}}\left(A\bar{B} + ABC\right)$.

Solution:

Given:

$$\overline{A + \overline{BC}}\left(A\bar{B} + ABC\right)$$

$$\overline{A + \overline{BC}}\left(A\bar{B} + ABC\right) = \bar{A}.BC\left(A\bar{B} + ABC\right)$$

$$= \bar{A}.BC.A\bar{B} + \bar{A}BC.ABC = 0$$

2. Let us prove that (a) $a + \bar{a}b = a + b$ and $\overline{xy}z + \bar{x}yz + x\bar{y} = \bar{x}z + x\bar{y}$.

Solution:

Given:

$$a + \bar{a}b = a + b$$

$$\overline{xy}z + \bar{x}yz + x\bar{y} = \bar{x}z + x\bar{y}$$

(a) $a + \bar{a}b = a + b$

$$a + \bar{a}b = a + ab + \bar{a}b = a + b\left(a + \bar{a}\right)(\because a + \bar{a} = 1)$$

$$= a + b$$

L.H.S. = R.H.S.

(b) $\bar{x}\bar{y}z + \bar{x}yz + x\bar{y} = \bar{x}z + x\bar{y}$

$$z\bar{x}\left(y + \bar{y}\right) + x\bar{y} = z\bar{x}\left(1\right) + x\bar{y} = z\bar{x} + x\bar{y}$$

L.H.S = R.H.S

2.3 Minimization of Switching Functions using K-Map upto 6 Variables

A Karnaugh map or K-map is a pictorial form of truth table in which the map diagram is made up of cells with each cell representing one min term or max term of the function. It provides a simple straight forward procedure for minimizing Boolean function.

Construction of a Karnaugh Map

An n-variable Karnaugh map has 2^n squares and each possible input is allocated a square. In the case of a min term Karnaugh map, '1' is placed in all those squares for which the output is '1' and '0' is placed in all those squares for which the output is '0'. 0s are omitted for simplicity. An 'X' is placed in squares corresponding to don't care conditions.

In the case of a max term Karnaugh map, a '1' is placed in all those squares for which the result is '0' and a '0' is placed for input entries corresponding to a '1' result. Then, 0s are omitted for simplicity and an 'X' is placed in squares corresponding to don't care conditions. The choice of terms identifying different rows and columns of a Karnaugh map is not unique for a given number of variables.

The only condition to be satisfied is that the designation of adjacent rows and adjacent columns should be the same except for one of the literals being complemented. Also, the extreme rows and extreme columns are considered adjacent. Some of the possible designation styles for 2, 3 and 4 variable min term, Karnaugh maps are shown in the figure below:

1. Two variable K-Map:

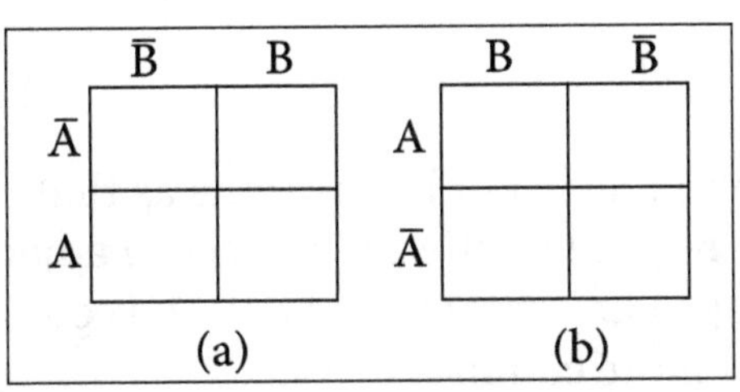

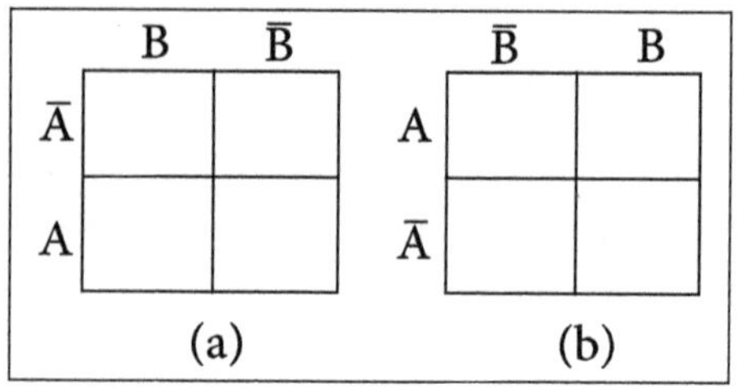

2. Three variable K-Map:

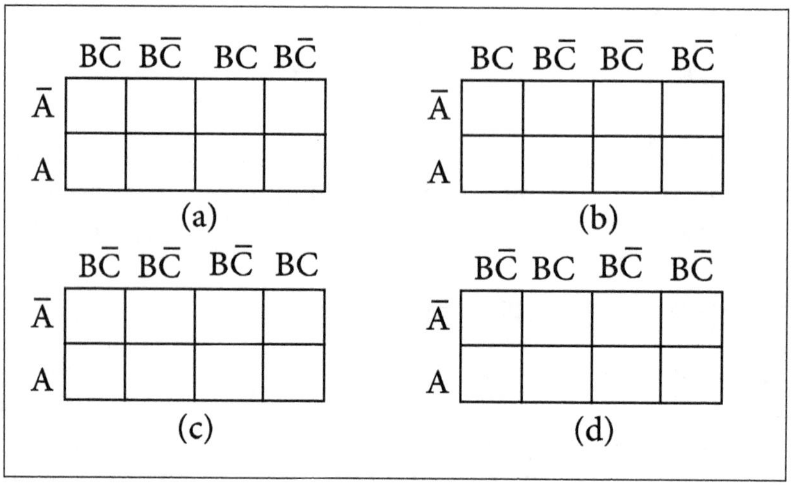

3. Four variable K-Map:

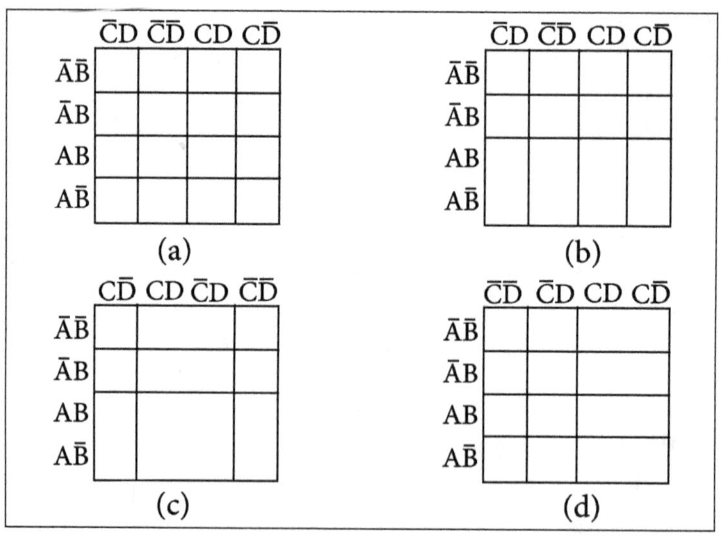

The style of row identification need not be the same as that of column identification as long as it meets the basic requirement with respect to adjacent terms. It is, however, accepted in practice to adopt a uniform style of row and column identification. Also, the style shown in the figure is more commonly used.

Having drawn the Karnaugh map, the next step is to form groups of 1s as per the subsequent guidelines:

- Each square containing a '1' must be considered at least once, although it can be considered as often as desired.
- The goal should be to account for all the marked squares in the minimum number of groups.
- The number of squares in a group should always be a power of 2, (i.e.,) groups can have 1,2, 4, 8, 16, squares.

4. Five variable K-Map:

A=0

BC \ DE	$\bar{D}\bar{E}$ 00	$\bar{D}E$ 01	DE 11	$D\bar{E}$ 10
$\bar{B}\bar{C}$ 00	0	1	3	2
$\bar{B}C$ 01	4	5	7	6
BC 11	12	13	15	14
$B\bar{C}$ 10	8	9	11	10

A=1

BC \ DE	$\bar{D}\bar{E}$ 00	$\bar{D}E$ 01	DE 11	$D\bar{E}$ 10
$\bar{B}\bar{C}$ 00	16	17	19	18
$\bar{B}C$ 01	20	21	23	22
BC 11	28	29	31	30
$B\bar{C}$ 10	24	25	27	26

5. Six variable K-Map:

$\bar{A}\,\bar{B}$

CD \ EF	$\bar{E}\bar{F}$ 00	$\bar{E}F$ 01	EF 11	$E\bar{F}$ 10
$\bar{C}\bar{D}$ 00	0	1	3	2
$\bar{C}D$ 01	4	5	7	6
CD 11	12	13	15	14
$C\bar{D}$ 10	8	9	11	10

$\bar{A}\,D$

CD \ EF	$\bar{E}\bar{F}$ 00	$\bar{E}F$ 01	EF 11	$E\bar{F}$ 10
$\bar{C}\bar{D}$ 00	16	17	19	18
$\bar{C}D$ 01	20	21	23	22
CD 11	28	29	31	30
$C\bar{D}$ 10	24	25	27	26

$A\,\bar{B}$

CD \ EF	$\bar{E}\bar{F}$ 00	$\bar{E}F$ 01	EF 11	$E\bar{F}$ 10
$\bar{C}\bar{D}$ 00	32	33	35	34
$\bar{C}D$ 01	36	37	39	38
CD 11	44	45	47	46
$C\bar{D}$ 10	40	41	43	42

$A\,B$

CD \ EF	$\bar{E}\bar{F}$ 00	$\bar{E}F$ 01	EF 11	$E\bar{F}$ 10
$\bar{C}\bar{D}$ 00	48	49	51	50
$\bar{C}D$ 01	52	53	55	54
CD 11	60	61	63	62
$C\bar{D}$ 10	56	57	59	58

Limitations of Karnaugh Map

- The map method is restricted in its capability because they are useful for simplifying only the Boolean expression represented in standard form.
- Generally it is limited to six variable map (i.e.,) more than six variables involving expressions are not reduced.

Problems

1. Let us simplify the following function using K-map and also implement the function using logic gates f(A, B, C) = π(0, 4, 6).

Solution:

Given:

$$f(A, B, C) = \pi(0, 4, 6)$$

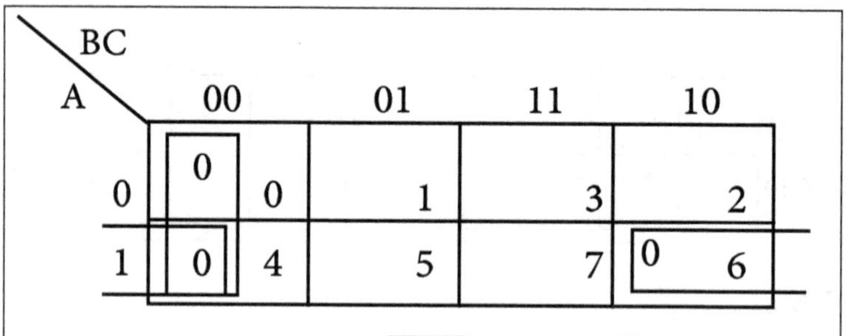

$$F = (\overline{B} + \overline{C})(A+B+\overline{C})$$

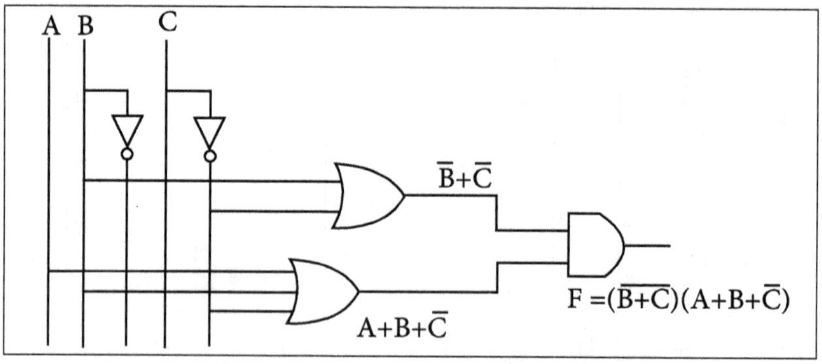

2. Let us reduce the following expression using K-map: F = m2 + m3 + m4 + m6 + m7 + m9 + m11 + m13.

Solution:

Given:

$$F = m2 + m3 + m4 + m6 + m7 + m9 + m11 + m13$$

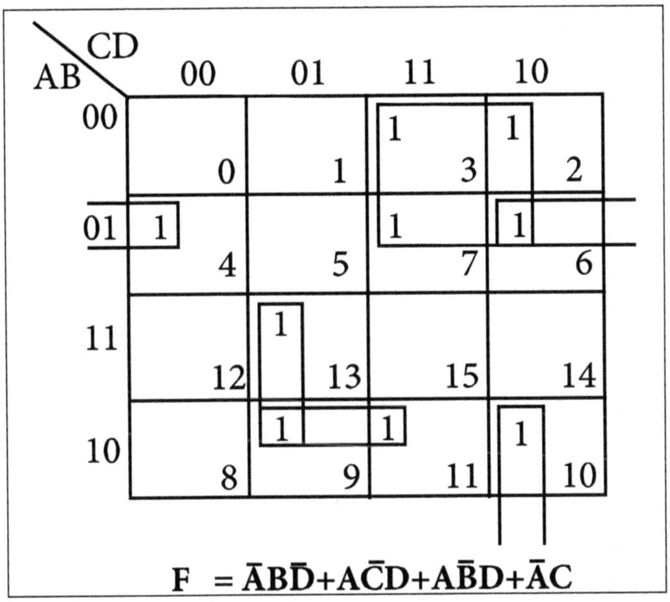

3. Let us reduce the following expression using K - map, f = x'y'z + w'xyz' + wxz + w'xyz.

Solution:

Given:

$$f = x'y'z + w'xyz' + wxz + w'xyz$$
$$= (w + w')\,x'y'z + w'xyz' + wx\,(y + y')\,z + w'xyz$$
$$f = wx'y'z + w'x'y'z + w'xyz + wxy'z + w'xyz$$
$$f = \Sigma m\,(9, 1, 6, 15, 13, 7)$$
$$= \Sigma m\,(1, 6, 7, 9, 13, 15)$$
$$f = x'y'z + wxz + w'xy$$

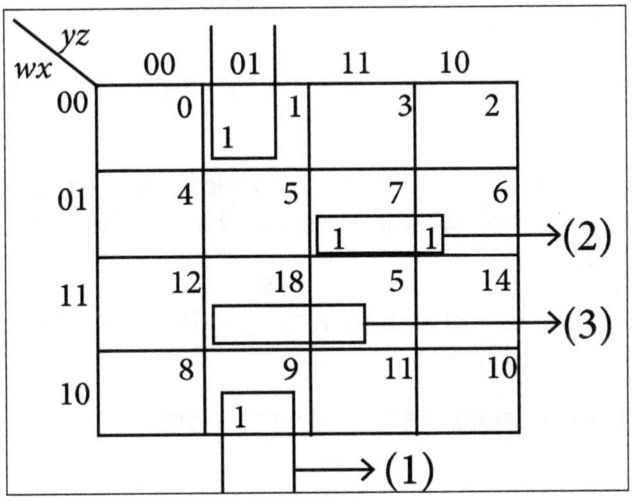

2.4 Tabular Minimization

A method involving an exhaustive tabular search method for the minimum expression to solve a Boolean equation for more variables is called as tabulation method.

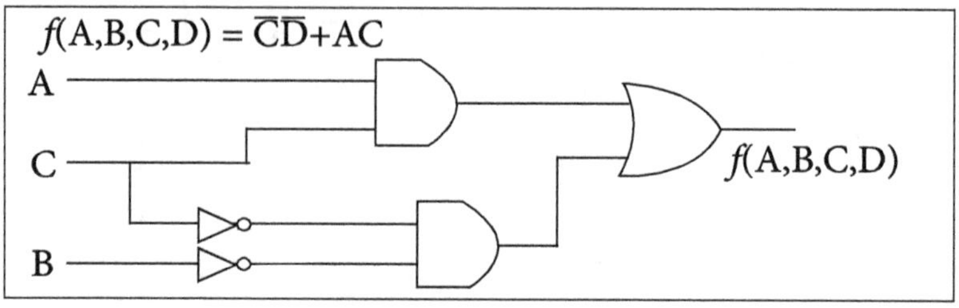

Simplified logic diagram.

The same procedure is applied for the other pairs of terms wherever such a reduction is possible. All these terms reduced by one literal are further examined if they can be reduced further. The process continues until the terms become irreducible. The irreducible terms are called prime implicants.

An optimum set of prime implicants that can account for all the original terms then constitutes the minimized expression. The technique can be applied equally well for minimizing sum-of-products and product-of-sums expressions and is particularly useful for Boolean functions having more than six variables as it can be mechanized and run on a computer.

On the other hand, the Karnaugh mapping method is a graphical method and becomes very cumbersome when the number of variables exceeds six. The step-by-step procedure for application of the tabular method for minimizing Boolean expressions, both sum-of-products and product-of-sums, is outlined as follows:

- The Boolean expression to be simplified is expanded, if it is not in expanded form.
- Different terms in the expression are divided into groups depending upon the number of 1s they have.
- True and complemented variables in a sum-of-products expression mean 1 and 0 respectively.
- The reverse is true in the case of a product-of-sums expression. The groups are then arranged, beginning with the group having the least number of 1s in its included terms.
- Terms within the same group are arranged in ascending order of the decimal numbers represented by these terms.

As an illustration, let us consider the below expression:

$$A.B.C + \bar{A}.B.C + A.\bar{B}.\bar{C} + A.\bar{B}.C + \bar{A}.\bar{B}.\bar{C}.$$

$\bar{A}.\bar{B}.\bar{C}$		000	First group
$A.\bar{B}.\bar{C}$	$\rightarrow$	100	Second group
$\bar{A}.B.C$		011	Third group
$A.\bar{B}.C$		101	
ABC		111	Fourth group

As another illustration, consider a product-of-sums expression given by:

$$\left(\bar{A}+\bar{B}+\bar{C}+\bar{D}\right).\left(\bar{A}+\bar{B}+\bar{C}+D\right).\left(\bar{A}+B+\bar{C}+D\right).\left(A+B+\bar{C}+\bar{D}\right).\left(A+B+C+D\right).$$

$$\left(A+\bar{B}+\bar{C}+\bar{D}.\left(A+\bar{B}+C+\bar{D}\right).$$

The formation of groups and the arrangement of terms within different groups for the product-of sums expression are as follows:

- It may be mentioned here that the Boolean expressions that we have considered above did not contain any optional terms.
- If there are any, they are also considered while forming groups. This completes the first table group to look for any possible matching and consequent reduction.

$A.B.C.D$		0000
–		
$A.B.\bar{C}.\bar{D}$		0011
$A.\bar{B}.C.\bar{D}$		0101
$\bar{A}.B.\bar{C}.D$	$\rightarrow$	1010
–		–
$A.\bar{B}.\bar{C}.\bar{D}$		0111
$\bar{A}.\bar{B}.\bar{C}.\bar{D}$		1110
$\bar{A}.\bar{B}.\bar{C}.\bar{D}$		–
–		–

The terms are considered as matched for all literals except for one match. The pairs of matched terms are replaced with a single term where the position of the unmatched literals is replaced with a dash (-).

These new terms formed as a result of the matching process find a place in the second table. The terms in the first table that do not find a match are called the prime implicants and are marked with an asterisk (*). The matched terms are ticked (✓).

Terms in the second group are compared with those in the third group to look for a possible match. Again, terms in the second group that do not find a match become the prime implicants. The process continues until we reach the last group. This completes the first round of matching. The terms resulting from the matching in the first round are recorded in the second table.

The next step is to perform matching operations in the second table. While comparing the terms for a match, it is important that a dash (-) is also treated like any other literal, that is, the dash signs also need to match. The process continues on to the third table, the fourth table and so on until the terms becomes irreducible any further.

An optimum selection of prime implicants to account for all the original terms constitutes the terms for the minimized expression. Although optional (also called don't care) terms are considered for matching, they do not have to be accounted for once prime implicants have been identified.

Problem

1. Let us simplify F(V, W, X, Y, Z) = ∑(0, 1, 8, 11, 12, 15, 20, 21, 22, 24, 29, 31) using tabulation method, which has the don't care conditions d (9, 18, 30).

Solution:

Given:

F(V, W, X, Y, Z) = ∑(0, 1, 8, 11, 12, 15, 20, 21, 22, 24, 29, 31)

Don't care conditions: d (9, 18, 30)

00000 (0)√	0000X (0,1)√	0X00X (0,1,8,9) 0X00X (0,8,1,9)
00001 (1)√ 01000 (8)√	0X000 (0,8)√	
01001 (9)√ 01100 (12)√ 10010 (18)√ 10100 (20)√ 11000 (24)√	0X001 (1,9)√ 0100X (8,9)√ 01X00 (8,12) X1000 (8,24)	
01011 (11)√ 10101 (21)√ 10110 (22)√	010X1 (9,11) 10X10 (18,22) 1010X (20,21) 1011X (20,22)	

01111 (15)√ 11101 (29)√ 11110 (30)√ 11111 (31)√	01X11 (11,15) 1X101 (21,29) 1X110 (22,30)	
	X11111 (15,31) 111X1 (29,31) 1111X (30,31)	

	0	1	8	9	11	12	15	18	20	21	22	24	29	30	31
0X00X	X	X	X	X											
(0,1,8,9)															
01X00			X			X									
(8,12)															
X1000			X									X			
(8,24)															
010X1				X	X										
(9,11)															
10X10								X			X				
(18,22)															
1010X									X	X					
(20,21)															
1011X									X		X				
(20,22)															
01X11					X		X								
(11,15)															
1X101										X				X	
(21,29)															
1X110							X								X
(22,30)															
X1111										X	X				
(15,31)															
111X1												X			X
(29,31)															
1111X														X	X
(30,31)															

F(V,W,X,Y,Z) = 0X00X + 01X00+ X1000 + 10X10 + 1010X + 01X11 + 1X110 + 111X1 + 1111X

$$= \overline{v}\,\overline{x}\,\overline{y} + \overline{v}\,w\,\overline{y}\,\overline{z} + w\,\overline{x}\,\overline{y}\,\overline{z} + v\overline{\omega}\,y\,\overline{z} + + v\overline{\omega}x\overline{y} + \overline{v}xy\overline{z} + v\omega wz + v\omega wz + v\omega xy$$

2.5 Problem Solving (Code-Converters using K-Map)

Code Conversion

There is a wide variety of binary codes used in digital systems. Some of these codes are binary-coded-decimal (BCD), Excess-3, Gray and so on. Many times it is required to convert one code to another.

Binary to BCD Convertor

BCD is binary coded decimal number, where each digit of a decimal number is respected by its equivalent binary number. That means, LSB of a decimal number is represented by its equivalent binary number and similarly other higher significant bits of decimal number are also represented by their equivalent binary numbers. For example, BCD Code 0001 (1) 0100 (4) of 14.

Let us design a 4bit binary to BCD code converter. As the 4 bit can represent 0 to 15, we can draw the conversion table as follows:

Binary code				BCD code				
D	C	B	A	B4	B3	B2	B1	B0
0	0	0	0	0	0	0	0	0
0	0	0	1	0	0	0	0	1
0	0	1	0	0	0	0	1	0
0	0	1	1	0	0	0	1	1
0	1	0	0	0	0	1	0	0
0	1	0	0	0	0	1	0	1
0	1	1	1	0	0	1	1	0
0	1	1	1	0	0	1	1	1
1	0	0	0	0	1	0	0	0
1	0	0	1	0	1	0	0	1
1	0	1	0	1	0	0	0	0
1	0	1	1	1	0	0	0	1
1	1	0	0	1	0	0	1	0
1	1	0	1	1	0	0	1	1
1	1	1	0	1	0	1	0	0
1	1	1	1	1	0	1	0	1

Here, B_4 bit represents MSB of decimal number and B_3, B_2, B_1, B_0 represents 4 bit binary equivalent of LSB of decimal number.

From, above conversion table, we can write SOP form for different bits of BCD code:

$$B_4 = \Sigma m(10, 11, 12, 13, 14, 15),\ B_3 = \Sigma m(8, 9),\ B_2 = \Sigma m(4, 5, 6, 7, 14, 15),$$

$$B_1 = \Sigma m(2, 3, 6, 7, 12, 13),\ B_0 = \Sigma m(1, 3, 5, 7, 9, 11, 13, 15)$$

K-map Simplification

For B_0

DC \ BA	00	01	11	10
00	0	1	1	0
01	0	1	1	0
11	0	1	1	0
10	0	1	1	0

$$B_0 = A$$

For B_1

DC \ BA	00	01	11	10
00	0	0	1	1
01	0	0	1	1
11	1	1	0	0
10	0	0	0	0

$$B_1 = DC\bar{B} + \bar{D}B$$

For B_2

DC \ BA	00	01	11	10
00	0	0	0	0
01	1	1	1	1
11	0	0	1	1
10	0	0	0	0

$$B_2 = \bar{D}C + CB$$

For B_3

DC \ BA	00	01	11	10
00	0	0	0	0
01	0	0	0	0
11	0	0	0	0
10	1	1	0	0

$$B_3 = D\bar{C}\bar{B}$$

For B_4

DC \ BA	00	01	11	10
00	0	0	0	0
01	0	0	0	0
11	1	1	1	1
10	0	0	1	1

$$B_4 = DC + DB$$

Logic Diagram:

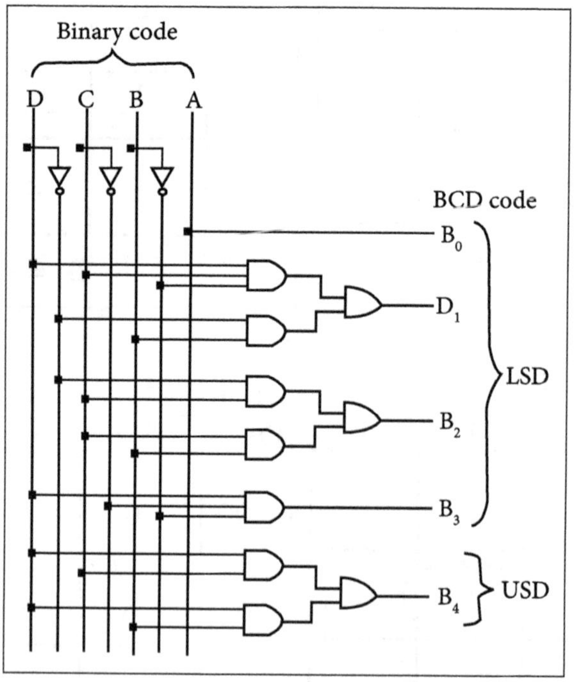

BCD to Binary Converter

A truth table matching BCD with the corresponding binary from 0 to 19 is given below. The decimal range 0 to 19 is chosen to illustrate the process but is easily extended to 39, 79, or 99, etc. This truth table will be used to design a BCD-to-binary module that can be cascaded to produce any size converter. It is possible to use a decimal range of 0 to 99 for this purpose, but the size of the module is considered too large to be of value for this example.

Let us see the truth table for BCD to binary converter:

B4	B3	B2	B1	Bo	E	D	C	B	A
0	0	0	0	0	0	0	0	0	0
0	0	0	1	1	0	0	0	0	1
0	0	0	0	0	0	0	0	1	0
0	0	0	1	1	0	0	0	1	1
0	0	1	0	0	0	0	1	0	0
0	0	1	0	1	0	0	1	0	1
0	0	1	1	0	0	0	1	1	0
0	0	1	1	1	0	0	1	1	1

0	1	0	0	0	0	1	0	0
0	1	0	0	1	0	1	0	1
1	0	0	0	0	0	1	1	0
1	0	0	0	1	0	1	1	1
1	0	0	1	0	0	1	0	0
1	0	0	1	1	0	1	0	1
1	0	1	0	0	0	1	1	0
1	0	1	0	1	0	1	1	1
1	0	1	1	0	1	0	0	0
1	0	1	1	1	1	0	0	1
1	1	0	0	0	1	0	1	0
1	1	0	0	1	1	0	1	1

K-Map Simplification

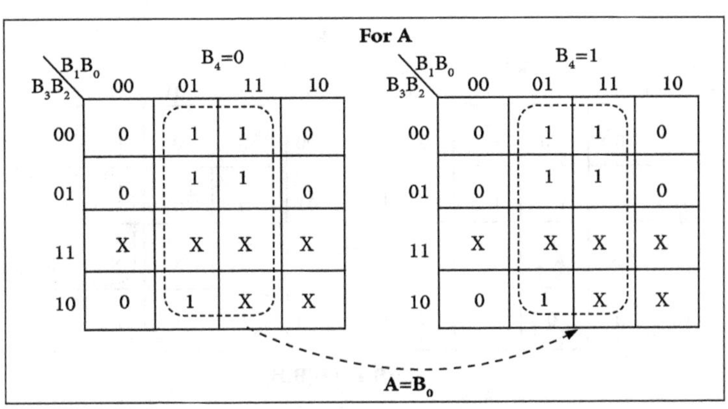

For B

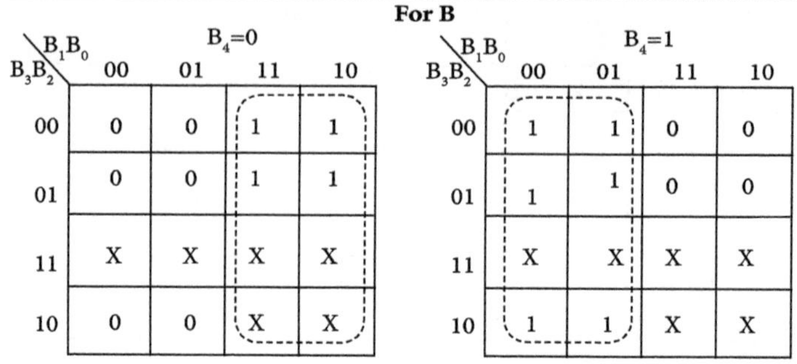

$$B=B_1\overline{B}_4+\overline{B}_1B_4$$
$$=B_1\oplus B_4$$

For C

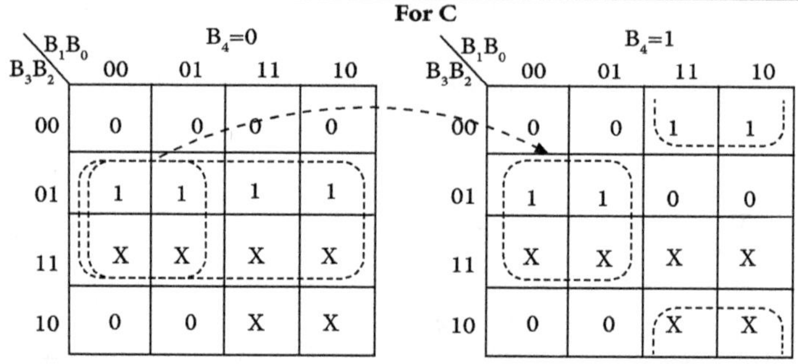

$$C=\overline{B}_4B_2+B_2\overline{B}_1+B_4\overline{B}_2B_1$$

For D

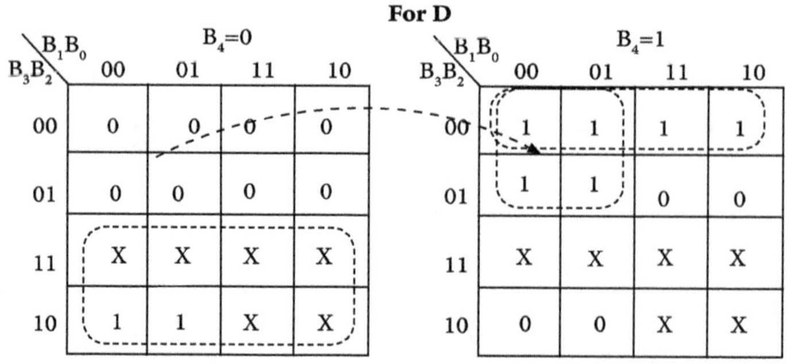

$$D=\overline{B}_4B_3+B_4\overline{B}_3\overline{B}_2+B_4\overline{B}_3\overline{B}_1$$

For E

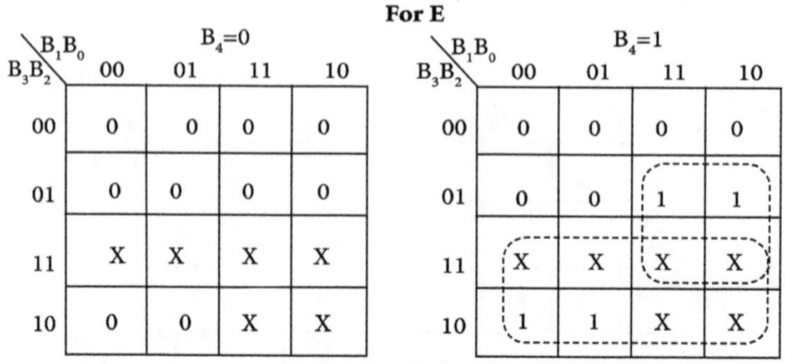

$$E=B_4B_3+B_4B_2B_1$$

Logic Diagram:

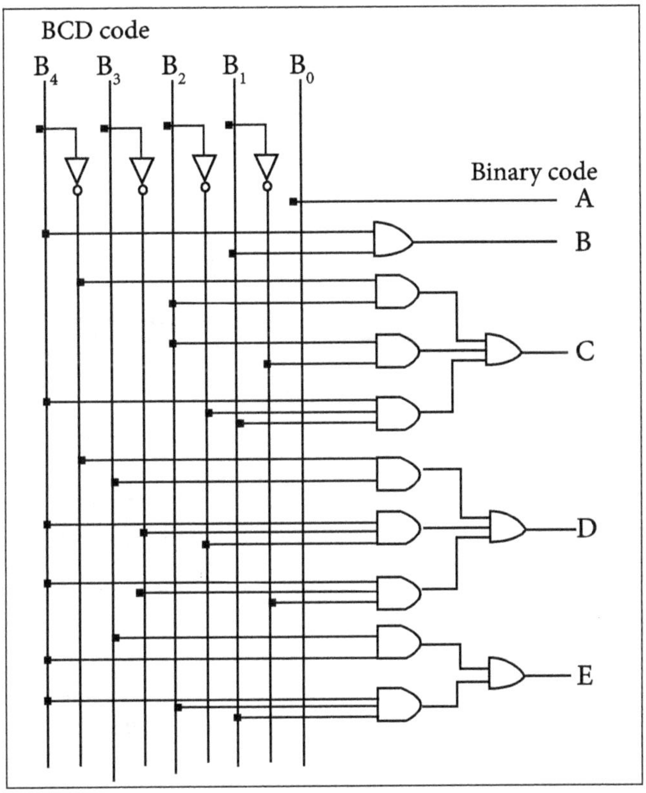

BCD to Excess-3 Converter

Excess-3 code is a modified form of a BCD number. The Excess-3 code can be derived from the natural BCD code by adding 3 to each coded number. For example, decimal 12 can be represented in BCD as 0001 0010. Now adding 3 to each digit we get Excess-3 code as 0100 0101 (12 in decimal). The truth table for BCD to Excess-3 code converter can be determined as shown in the below table.

Decimal	B3	B2	B1	Bo	E3	E2	E1	Eo
0	0	0	0	0	0	0	1	1
1	0	0	0	1	0	1	0	0
2	0	0	1	0	0	1	0	0
3	0	0	1	1	0	1	1	0
4	0	1	0	0	0	1	1	1
5	0	1	0	1	1	0	0	0
6	0	1	1	0	1	0	0	1
7	0	1	1	1	1	0	1	0
8	1	0	0	0	1	0	1	1
9	1	0	0	1	1	1	0	0

K-Map Simplification

For E_3

B_3B_2 \ B_1B_0	00	01	11	10
00	0	0	0	0
01	0	1	1	1
11	X	X	X	X
10	1	1	X	X

$\therefore E_3 B_3 + B_2(B_0+B_1)$

For E_2

B_3B_2 \ B_1B_0	00	01	11	10
00	0	1	1	1
01	1	0	0	0
11	X	X	X	X
10	0	1	X	X

$\therefore E_2 = B_2\bar{B}_1\bar{B}_0 + \bar{B}_2(B_0 + B_1)$

For E_1

B_3B_2 \ B_1B_0	00	01	11	10
00	1	0	1	0
01	1	0	1	0
11	X	X	X	X
10	1	0	X	X

$E_1 = \bar{B}_1\bar{B}_0 + B_1B_0$
$= B_1 \oplus B_0$

For E_0

B_3B_2 \ B_1B_0	00	01	11	10
00	1	0	0	1
01	1	0	0	1
11	X	X	X	X
10	1	0	X	X

$E_0 = \bar{B}_0$

Logic Diagram:

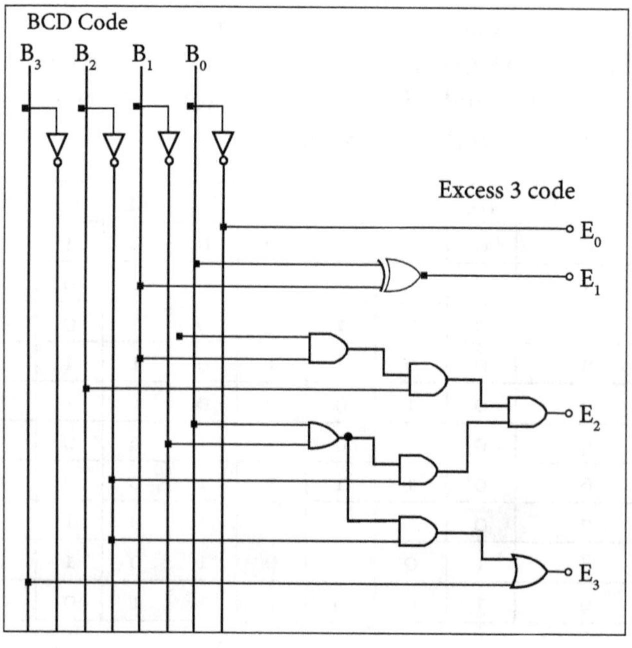

Excess-3 to BCD Code Converter

- BCD can be derived from Excess-3 code by reversing the process used in conversion of BCD to Excess-3 code which is to subtract 0011 or 0011 0011 from given BCD number.
- For example, Decimal number 12 is represented as 0100 0101 in Excess-3 code. If we subtract 3 (0011 0011) from given Excess-3 code, then the corresponding BCD is 0001 0010.
- Write the truth table relating Excess-3 and BCD.
- Write Karnaugh-Map for each output and obtain simplified expression.
- Implement Excess-3 code to BCD converter using Logic Gates.
- Implement Logic Gates' circuit in PLC using Ladder Diagram programming language.

The truth table for Excess-3 to BCD Code Converter is shown in the below table:

E_3	E_2	E_1	E_0	B_3	B_2	B_1	B_0
0	0	1	1	0	0	0	1
0	1	0	0	0	0	0	0
0	1	0	1	0	0	1	0
0	1	1	0	0	0	1	1
0	1	1	1	0	1	0	1
1	0	0	0	0	1	0	0
1	0	0	1	0	1	1	0
1	0	1	0	0	1	1	1
1	0	1	1	1	0	0	1
1	1	0	0	1	0	0	0

K-Map Simplification

For B_0

E_3E_2 \ E_1E_0	00	01	11	10
00	X	X	0	X
01	1	0	0	1
11	1	X	X	X
10	1	0	0	1

$B_0=\overline{E}_0$

For B_1

E_3E_2 \ E_1E_0	00	01	11	10
00	X	X	0	X
01	0	1	0	1
11	0	X	X	X
10	0	1	0	1

$B_1=\overline{E}_1 E_0+E_1\overline{E}_0$
$=E_1 \oplus E_0$

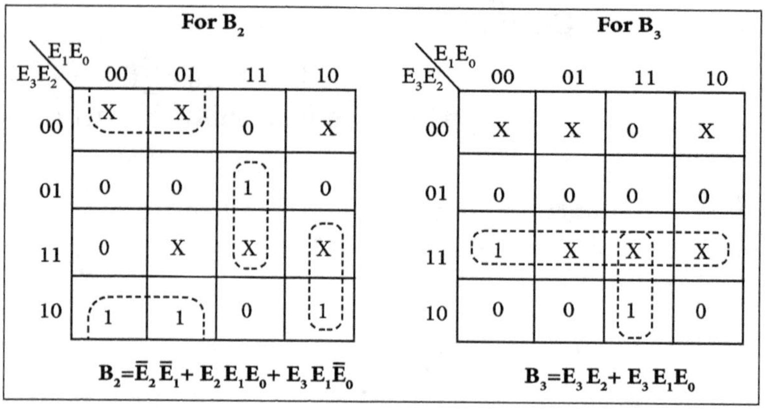

Logic Diagram:

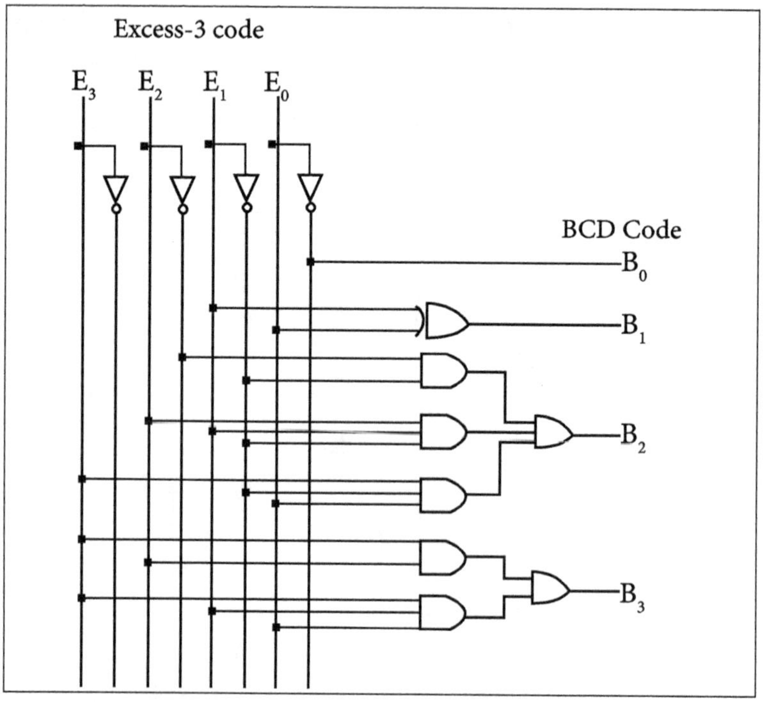

Binary to Gray Code Converter

The Gray code is often used in digital systems because it has the advantage that only one bit in the numerical representation changes between successive numbers. The below table shows decimal and Binary codes and corresponding Gray code.

Truth Table

Decimal	Binary Code				Gray Code			
	D	C	B	A	G3	G2	G1	Go
0	0	0	0	0	0	0	0	0

1	0	0	0	1	0	0	0	1
2	0	0	1	0	0	0	1	1
3	0	0	1	1	0	0	1	0
4	0	1	0	0	0	1	1	0
5	0	1	0	1	0	1	1	1
6	0	1	1	0	0	1	0	1
7	0	1	1	1	0	1	0	0
8	1	0	0	0	1	1	0	0
9	1	0	0	1	1	1	0	1
10	1	0	1	0	1	1	1	1
11	1	0	1	1	1	1	1	0
12	1	1	0	0	1	0	1	0
13	1	1	0	1	1	0	1	1
14	1	1	1	0	1	0	0	1
15	1	1	1	1	1	0	0	0

K-Map Simplification

For G_0

DC \ BA	00	01	11	10
00	0	1	0	1
01	0	1	0	1
11	0	1	0	1
10	0	1	0	1

$$G_0 = \bar{B}A + B\bar{A}$$
$$= B \oplus A$$

For G_1

DC \ BA	00	01	11	10
00	0	0	1	1
01	1	1	0	0
11	1	1	0	0
10	0	0	1	1

$$G_1 = C\bar{B} + \bar{C}B$$
$$= C \oplus B$$

For G_2

DC \ BA	00	01	11	10
00	0	0	0	0
01	1	1	1	1
11	0	0	0	0
10	1	1	1	1

$$G_2 = \bar{D}C + D\bar{C}$$
$$= D \oplus C$$

For G_3

DC \ BA	00	01	11	10
00	0	0	0	0
01	0	0	0	0
11	1	1	1	1
10	1	1	1	1

$$G_3 = D$$

Logic Diagram:

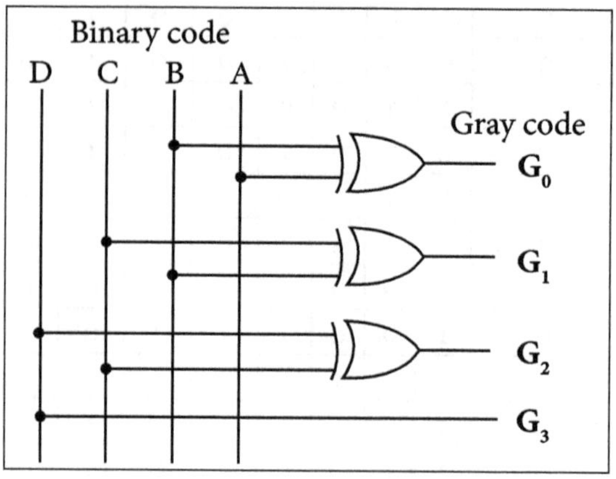

Gray Code to Binary Code Converter

In gray to binary code converter, input is a multiplies gray code and output is its equivalent binary code.

Let us consider a 4 bit gray to binary code converter. To design a 4 bit gray to binary code converter, we first have to draw a conversion table:

Gray Code				Binary Code			
G3	G2	G1	Go	D	C	B	A
0	0	0	0	0	0	0	0
0	0	0	1	0	0	0	1
0	0	1	1	0	0	1	0
0	0	1	0	0	0	1	1
0	1	1	0	0	1	0	0
0	1	1	1	0	1	0	1
0	1	0	1	0	1	1	0
0	1	0	0	0	1	1	1
1	1	0	0	1	0	0	0
1	1	0	1	1	0	0	1
1	1	1	1	1	0	1	0
1	1	1	0	1	0	1	1
1	0	1	0	1	1	0	0
1	0	1	1	1	1	0	1
1	0	0	1	1	1	1	0
1	0	0	0	1	1	1	1

K-Map Simplification

For A

G_3G_2 \ G_1G_0	00	01	11	10
00	0	1	0	1
01	1	0	1	0
11	0	1	0	1
10	1	0	1	0

For B

G_3G_2 \ G_1G_0	00	01	11	10
00	0	0	1	1
01	1	1	0	0
11	0	0	1	1
10	1	1	0	0

$$A=\left(\bar{G}_3G_2+G_3\bar{G}_2\right)\bar{G}_1\bar{G}_0+\left(\bar{G}_3\bar{G}_2+G_3\bar{G}_2\right)\bar{G}_1G_0$$
$$+\left(\bar{G}_3G_2+G_3\bar{G}_2\right)G_1G_0+\left(\bar{G}_3\bar{G}_2+G_3G_2\right)G_1\bar{G}_0$$

$$=\left(G_3\oplus G\right)\bar{G}_1G_0+\left(G_3\odot G_2\right)\bar{G}_1G_0+\left(G_3\oplus G_2\right)G_3G_0+\left(G_3\odot G_2\right)\bar{G}_1G_0$$
$$=\left(G_3\oplus G_2\right)\left(\bar{G}_1\bar{G}_0+G_1G_0\right)+\left(G_3\odot G_2\right)\left(\bar{G}_1G_0+G_1\bar{G}_0\right)$$
$$=\left(G_3\oplus G_2\right)\left(G_1\odot G_0\right)+\left(G_3\odot G_2\right)\left(G_1\oplus G_0\right)$$
$$=\left(G_3\oplus G_2\right)\left(\overline{G_1\oplus G_0}\right)+\left(\overline{G_3\oplus G_2}\right)\left(G_1\oplus G_0\right)$$
$$=\left(G_3\oplus G_2\right)\oplus\left(G_1\oplus G_0\right)$$

$$B=\left(\bar{G}_3\bar{G}_2+G_3G_2\right)G_1+\left(\bar{G}_3G_2+G_3\bar{G}_2\right)\bar{G}_1$$
$$=\left(G_3\odot G_2\right)G_1+\left(G_3\oplus G_2\right)\bar{G}_1=\left(\overline{G_3\oplus G_2}\right)G_1+\left(G_3\oplus G_2\right)\bar{G}_1$$
$$=G_3\oplus G_2\oplus G_1$$

For C

G_3G_2 \ G_1G_0	00	01	11	10
00	0	1	0	1
01	1	0	1	0
11	0	1	0	1
10	1	0	1	0

$C=\bar{G}_3G_2+G_3\bar{G}_2$
$=G_3\oplus G_2$

For D

G_3G_2 \ G_1G_0	00	01	11	10
00	0	0	0	0
01	0	0	0	0
11	1	1	1	1
10	1	1	1	1

$D=G_3$

Logic Diagram:

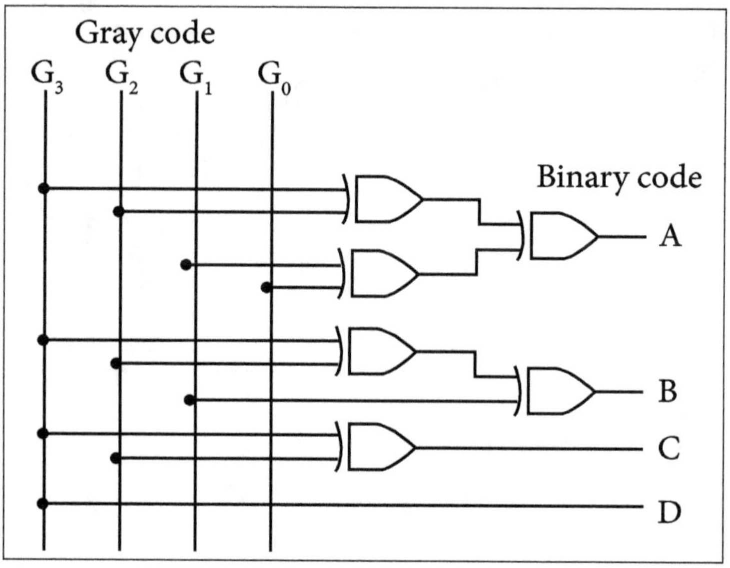

BCD to Gray Code Converter

The logical circuit which converts binary code to equivalent gray code is known as binary to gray code converter. The gray code is a non-weighted code. The successive gray code differs in one bit position only that means it is a unit distance code. It is also referred as cyclic code. It is not suitable for arithmetic operations. It is the most popular of the unit distance codes. It is also a reflective code. An n-bit Gray code can be obtained by reflecting an n-1 bit code about an axis after 2^{n-1} rows, and putting the MSB of 0 above the axis and the MSB of 1 below the axis. The 4 bits binary to gray code conversion table is given below:

BCD Code				Gray Code			
B3	B2	B1	B0	G3	G2	G1	G0
0	0	0	0	0	0	0	0
0	0	0	1	0	0	0	1
0	0	1	0	0	0	1	1
0	0	1	1	0	0	1	0
0	1	0	0	0	1	1	0
0	1	0	1	0	1	1	1
0	1	1	0	0	1	0	1
0	1	1	1	0	1	0	0
1	0	0	0	1	1	0	0
1	0	0	1	1	1	0	1

For G_0

$B_3B_2 \backslash B_1B_0$	00	01	11	10
00	0	1	0	1
01	0	1	0	1
11	X	X	X	X
10	0	1	X	X

$$G_0 = \bar{B}_1B_0 + B_1\bar{B}_0$$
$$= B_1 \oplus B_0$$

For G_1

$B_3B_2 \backslash B_1B_0$	00	01	11	10
00	0	0	1	1
01	1	1	0	0
11	X	X	X	X
10	0	0	X	X

$$G_1 = B_2\bar{B}_1 + \bar{B}_2B_1$$
$$= B_2 \oplus B_1$$

K-Map Simplification

For G_2

$B_3B_2 \backslash B_1B_0$	00	01	11	10
00	0	0	0	0
01	1	1	1	1
11	X	X	X	X
10	1	1	X	X

$$G_2 = \bar{B}_2 + B_3$$

For G_3

$B_3B_2 \backslash B_1B_0$	00	01	11	10
00	0	0	0	0
01	0	0	0	0
11	X	X	X	X
10	1	1	X	X

$$G_3 = B_3$$

Logic Diagram:

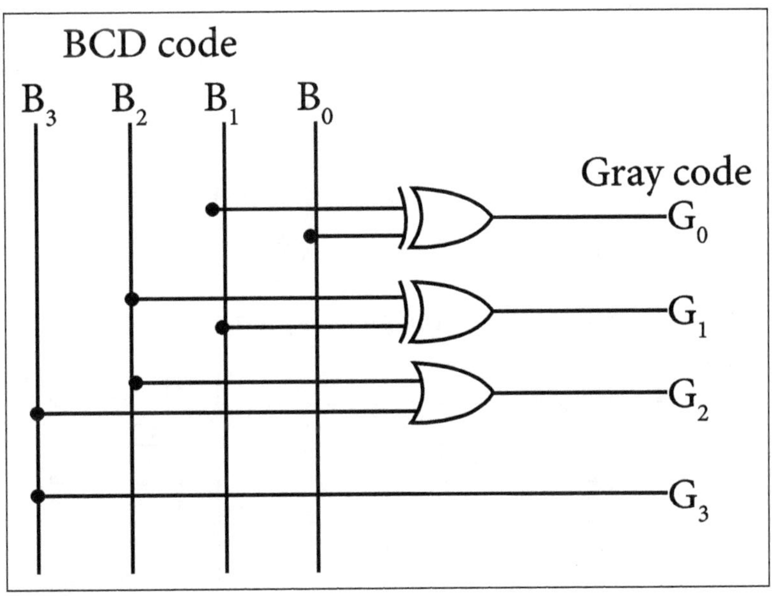

Problems

1. Simplify the given 2-variable Boolean equation by using K-map.

$$F = XY' + X'Y + X'Y'$$

First, let's construct the truth table for the given equation,

X	Y	F
0	0	0
0	1	1
1	0	1
1	1	1

We put 1 at the output terms given in equation.

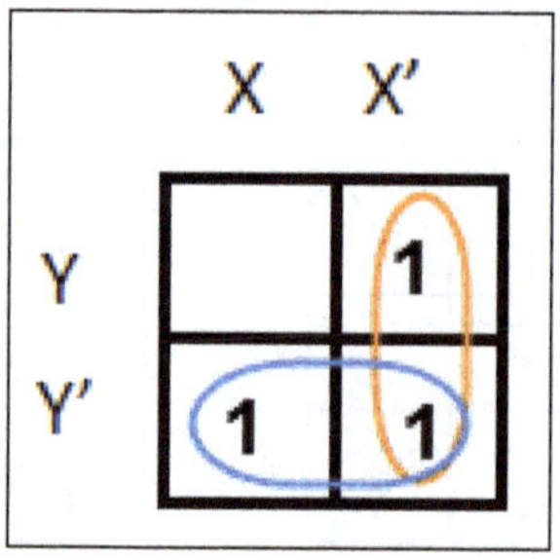

In this K-map, we can create 2 groups by following the rules for grouping, one is by combining (X',Y) and (X',Y') terms and the other is by combining (X,Y') and (X',Y') terms. Here the lower right cell is used in both groups.

After grouping the variables, the next step is determining the minimized expression.

By reducing each group, we obtain a conjunction of the minimized expression such as by taking out the common terms from two groups, i.e., X' and Y'. So the reduced equation will be X' +Y'.

2. Simplify the given 3-variable Boolean equation by using k-map.

$$F = X'YZ + X'Y'Z + XYZ' + X'Y'Z' + XYZ + XY'Z'$$

First, let's construct the truth table for the given equation,

x	y	z	F
0	0	0	1
0	0	1	1
0	1	0	0
0	1	1	1
1	0	0	1

1	0	1	1
1	1	0	1
1	1	1	0

We put 1 at the output terms given in equation.

There are 8 cells (23) in the 3-variable k-map.

The largest group size will be 8 but we can also form the groups of size 4 and size 2, by possibility. In the 3 variable Karnaugh map, we consider the left most column of the k-map as the adjacent column of rightmost column. So the size 4 group is formed as shown below:

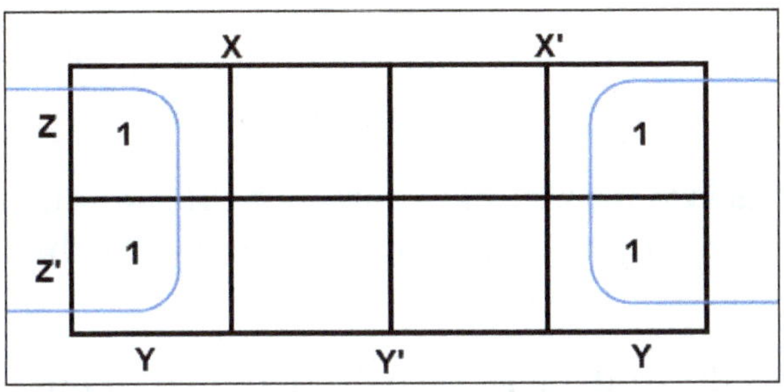

And in both the terms, we have 'Y' in common. So the group of size 4 is reduced as the conjunction Y. To consume every cell which has 1 in it, we group the rest of cells to form size 2 group, as shown below:

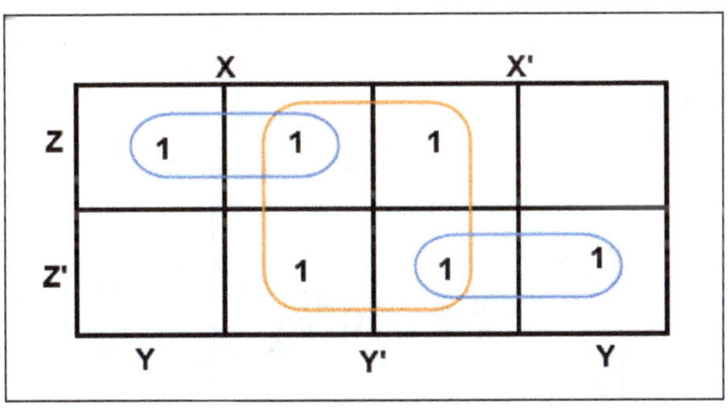

The 2 size group has no common variables, so they are written with their variables and its conjugates. So the reduced equation will be XZ' + Y' + X'Z. In this equation, no further minimization is possible.

3

Combinational Logic Circuits Design

3.1 Design of Half Adder and Full Adder

A logic circuit which performs the addition of two bits is a half adder. A half-adder is an arithmetic circuit block which can be used to add two bits.

Such a circuit thus has two inputs that represent the two bits to be added and two outputs with one which produces the SUM output and the other producing the CARRY.

The below figure shows the truth table of a half adder that shows all possible input combinations and the corresponding outputs.

The Boolean expressions for the outputs SUM and CARRY are shown in the equations given below as:

$$\text{SUM}\,S = A.\overline{B} + \overline{A}.B$$
$$\text{CARRY}\,C = A.B$$

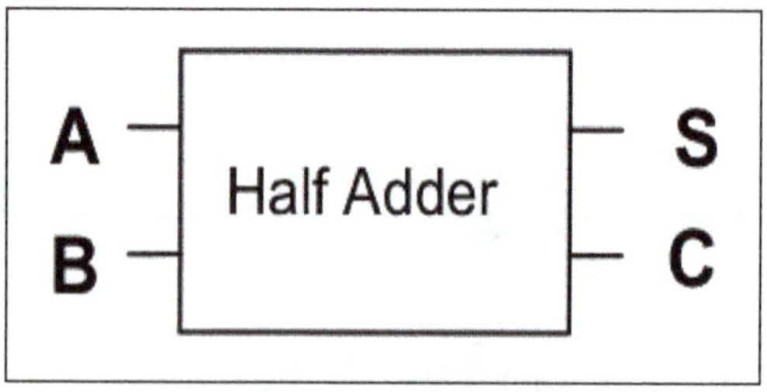

Block diagram of half adder.

Truth Table

A	B	S	C
0	0	0	0
0	1	1	0
1	0	1	0
1	1	0	1

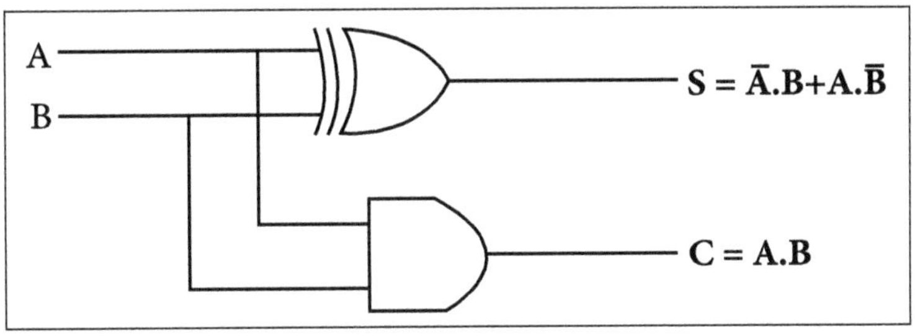

Half adder.

Full Adder

The circuit that performs the addition of three bits is a full adder.

Implementation of a Full Adder Using Half Adders

The first two inputs are A and B and the third input is an input carry as C-IN. The output carry is designated as C-OUT and the normal output is designated as S which is SUM.

Full adder accepts three inputs and it generates two outputs.

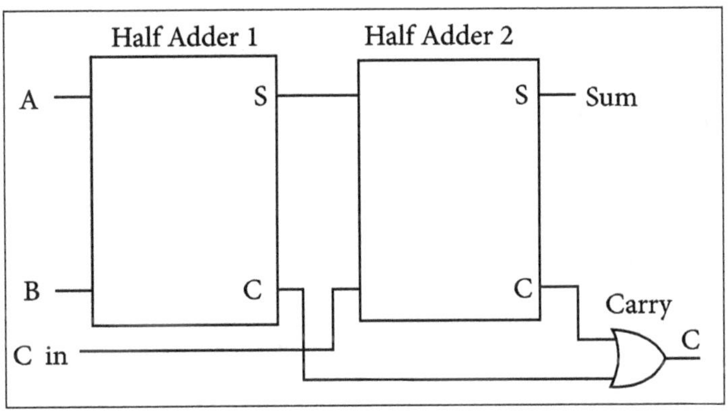

The resultant output will be named as C_{out}.

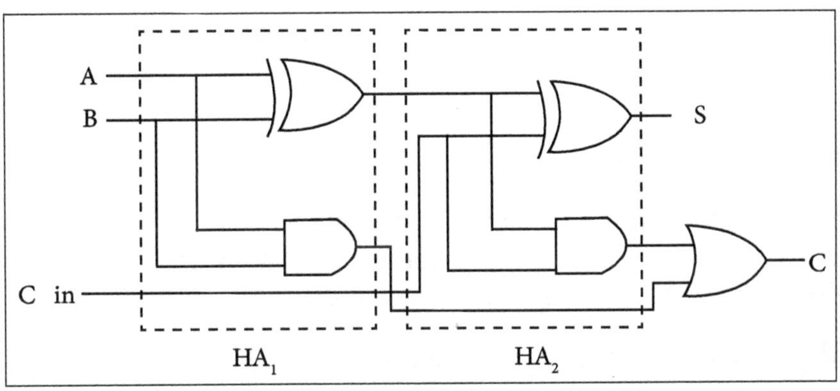

Implementation of Full Adder using half adders.

Truth Table

Input			Output	
A	B	Cin	S	C
0	0	0	0	0
0	0	1	1	0
0	1	0	1	0
0	1	1	0	1
1	0	0	1	0
1	0	1	0	1
1	1	0	0	1
1	1	1	1	1

3.2 Half Subtractor and Full Subtractor: Applications of Full Adders

Half Subtractor

A half-subtractor is a combinational circuit which is used to subtract one binary digit from another to produce a DIFFERENCE output and a BORROW output.

The BORROW output specifies whether a = 1 has been borrowed to perform the subtraction. The Boolean expressions for the two outputs are given by the equations.

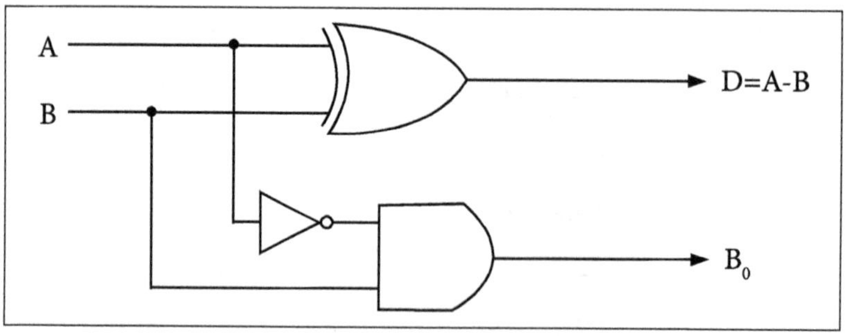

Half subtractor.

Full Subtractor

A full subtractor performs the subtraction function on two bits such as a minuend and a subtrahend and also takes into consideration whether a = 1 has already been borrowed by the previous adjacent lower minuend bit or not.

As a result there are three bits to be handled at the input of a full subtractor namely the two bits to be subtracted and a borrow bit that is designated as B_{in}.

There are two outputs namely the DIFFERENCE output D and a BORROW output B_O.

The BORROW output bit tells whether the minuend bit needs to borrow a =1 from the next possible higher minuend bit.

The below figure shows the truth table of a full subtractor. The Boolean expressions are given by the equations for two output variables:

$$D = \bar{A}.\bar{B}.B_{in} + \bar{A}.B.\bar{B}_{in} + A.\bar{B}.\bar{B}_{in} + A.B.B_{in}$$

$$B_o = \bar{A}.\bar{B}.B_{in} + \bar{A}.B.\bar{B}_{in} + \bar{A}.B.B_{in} + A.B.B_{in}$$

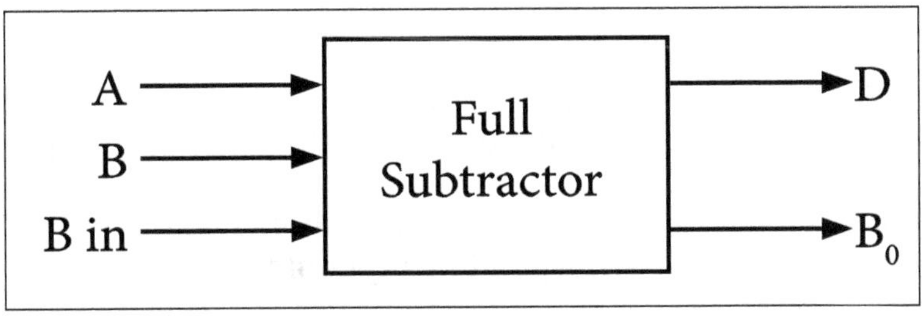

Block diagram of full subtractor.

Truth Table

Minuend (A)	Subtrahend (B)	Borrow In (Bin)	Difference (D)	Borrow out(Bo)
0	0	0	0	0
0	0	1	1	1
0	1	0	1	1
0	1	1	0	1
1	0	0	1	0
1	0	1	0	0
1	1	0	0	0
1	1	1	1	1

3.2.1 Applications of Full Adders

- Full adder reduces the circuit complexity.
- This is used to construct a ripple carry counter to add n-bit numbers. Thus, it is used in the Arithmetic and Logic Unit also.
- This can be used in Processor chip such as Snapdragon, Intel Pentium or Exynous for CPU part which consists of ALU. Therefore, this block is used to make operations such as Add, Subtract, Multiply etc.

- A one-bit full adder adds three one-bit numbers, often written as A, B and C_{in} where A and B are the operands and C_{in} is a bit carried in from the previous less significant stage.
- A full adder adds the binary numbers and accounts for values carried in as well as out.

3.3 4-Bit Binary Subtractor: Adder-Subtractor Circuit

The subtraction of binary numbers can be done most conveniently by the means of complements.

A subtraction A-B can be done by taking the 2's complement of B and adding it to A. A 2's complement can be obtained by taking 1 's complement and adding one to the LSB.

The circuit for subtracting the A -B consists of a parallel adder with inverters placed between each data input B and corresponding input of full adder as:

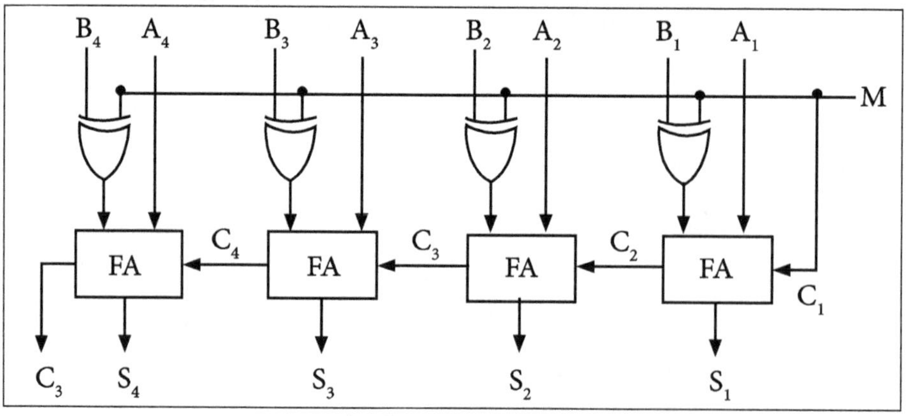

4-bit binary subtractor.

When M = 1, then the carry input $C_1 = 1$ and the operation is A plus 1's complements of B plus one. This equals A plus two's complement of B.

For signed numbers, the result is A -B provided there is no overflow.

For unsigned numbers, this gives A -B if $A > B$ or 2 's complement of $B - A$ if $A < B$.

3.3.1 Adder-subtractor Circuit

Serial Adder

Serial adder is defined as it would accept bit by bit input of the n-bit numbers and there is a bit by bit output of the n-bit Sum. Here, we would be required one full adder and a memory element. Hence, this requires lesser hardware.

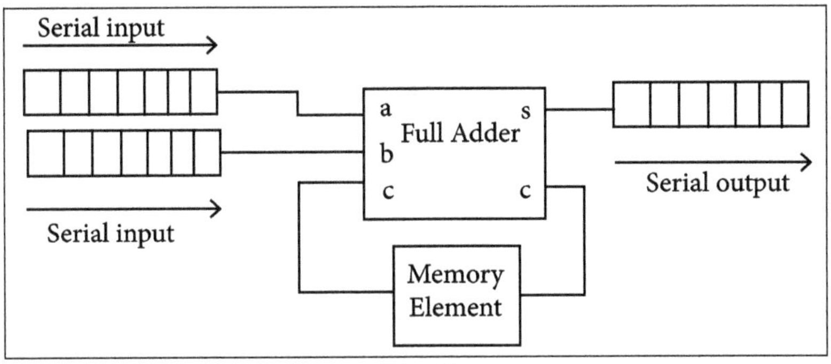

Serial adder.

Serial Subtractor

In serial subtractor, we have Input number coming bit by bit and output comes bit by bit and the final borrow at the end. The circuit for serial subtractor as follows:

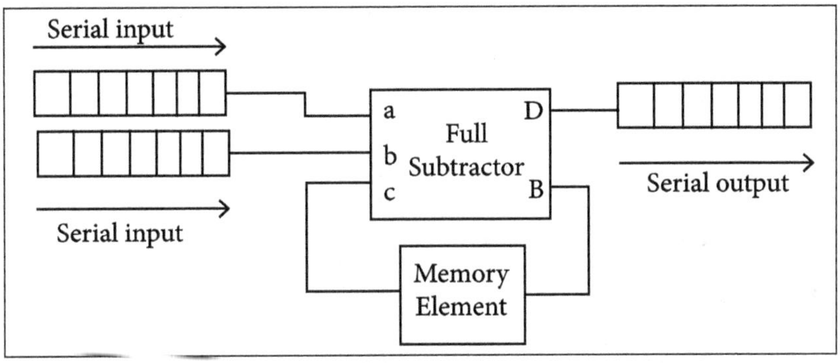

Serial subtracter.

Parallel Binary Adder

Parallel adders are digital circuits which compute the addition of variable binary strings with equivalent or different sizes in parallel.

The schematic diagram of a parallel adder is shown in the below figure:

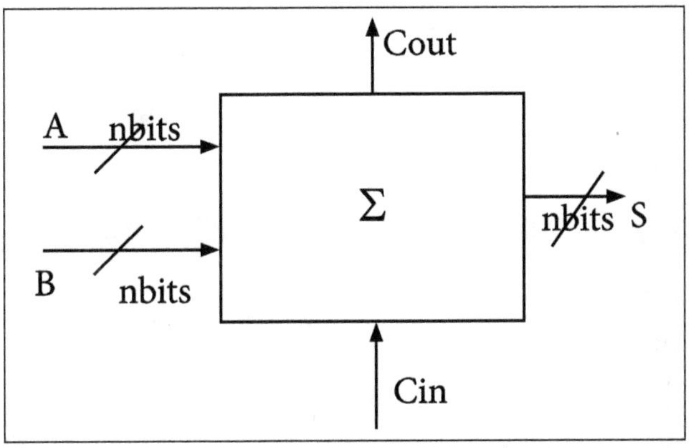

Parallel adder.

4 Bit Binary Ripple Carry Adder

Let the truth table for 4 bit binary binary full adder will be as:

A	B	C	D	S	CY
0	0	0	0	0	0
0	0	0	1	1	0
0	0	1	0	1	0
0	0	1	1	0	1
0	1	0	0	1	0
0	1	0	1	0	1
0	1	1	0	0	1
0	1	1	1	1	1
1	0	0	0	1	0
1	0	0	1	0	1
1	0	1	0	0	1
1	0	1	1	1	1
1	1	0	0	0	1
1	1	0	1	1	1
1	1	1	0	1	1
1	1	1	1	0	1

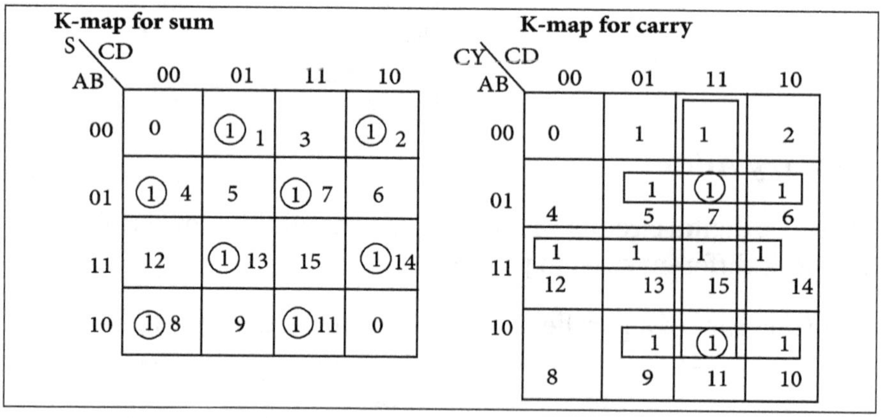

$$S = \bar{A}\bar{B}\bar{C}D + \bar{A}\bar{B}C\bar{D} + \bar{A}B\bar{C}D + AB\bar{C}D + ABC\bar{D} + A\bar{B}\bar{C}\bar{D} + A\bar{B}CD$$

$$= \bar{A}\bar{B}\left[\bar{C}D + C\bar{D}\right] + \bar{A}B\left[\bar{C}D + CD\right]AB\left[\bar{C}D + C\bar{D}\right] + A\bar{B}\left[\bar{C}D + CD\right]$$

$$= \bar{A}\bar{B} + [C \oplus D] + \bar{A}B\left[\bar{C}D + CD\right] + AB[C \oplus D] + A\bar{B} + [C \oplus D]$$

$$S = A \oplus B \oplus C \oplus D + A \oplus B \oplus C \oplus D$$

$$CY = CD + \bar{A}BD + \bar{A}BC + AB + A\bar{B}D + A\bar{B}C$$

$$= CD + D\left[\bar{A}B + A\bar{B}\right] + C\left(\bar{A}B + \bar{B}\right) + A\bar{B}C$$

$$= AB + CD + A \oplus B \oplus C \oplus D$$

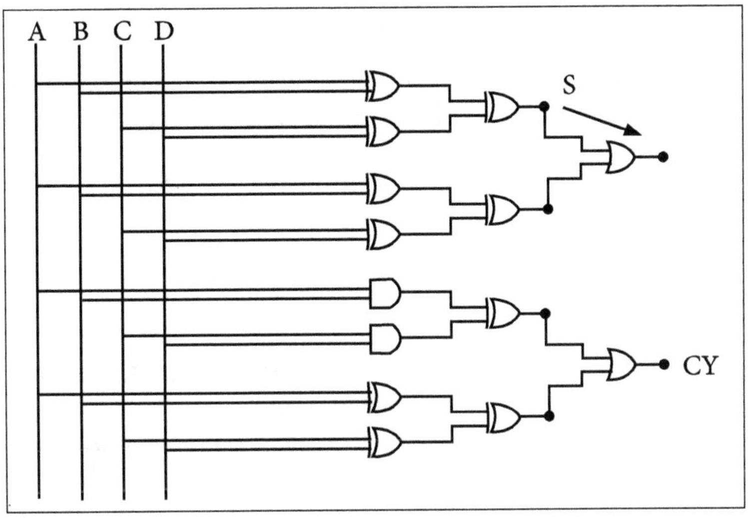

Parallel Binary Subtractor

To perform the subtraction of binary numbers with more than one bit is performed through the parallel subtractors.

This can be designed in several ways including the combination of half and full subtractors, all the full subtractors, all the full adders with subtrahend complement input, etc.

The below figure shows a 4 bit parallel binary subtractor which is formed by connecting one half subtractor and three full subtractors.

In this subtractor, 4 bit minuend $A_3A_2A_1A_0$ is being subtracted by a 4 bit subtrahend $B_3B_2B_1B_0$ and this gives the difference output $D_3D_2D_1D_0$.

The borrow output of the each subtractor is being connected as the borrow input to the next preceding subtractor.

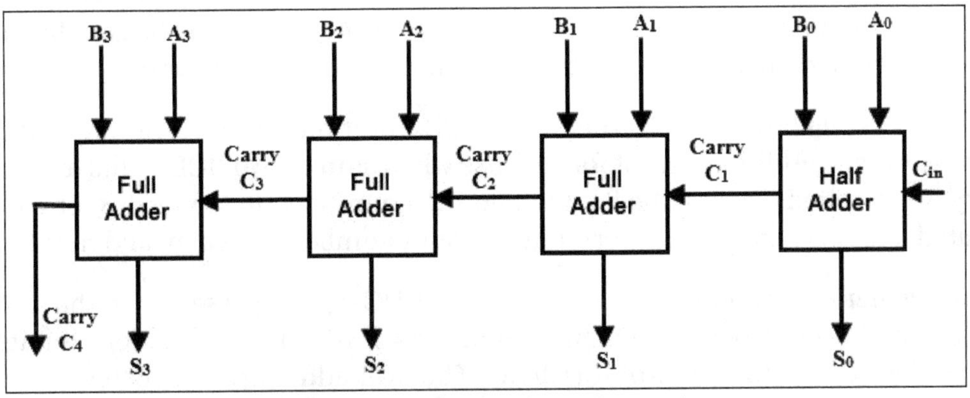

Full adders with subtrahend complement input.

It is also possible to design a 4 bit parallel subtractor as shown in the below figure.

The subtraction operation is performed by considering the principle that the addition of minuend and the complement of the subtrahend is equivalent to the subtraction process.

Therefore, the subtraction of A by B is obtained by taking 2's complement of B and adding it to A.

2's complement of B can be obtained by taking 1's complement and adding 1 to the least significant pair of bits.

Thus, in this circuit 1's complement of B is obtained with the inverters (NOT gate) and 1 can be added to the sum through the input carry.

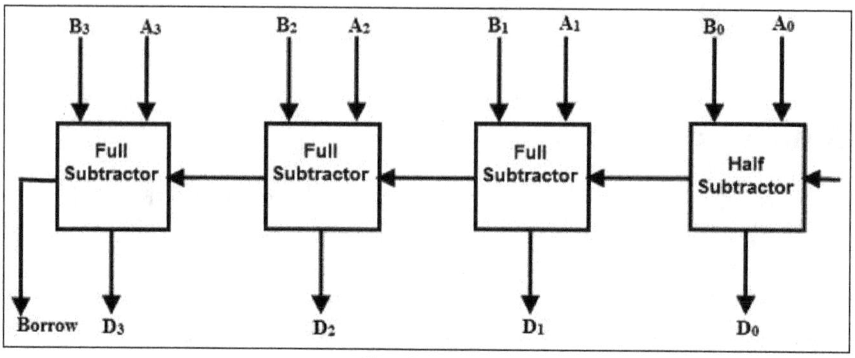

Full subtractors.

3.4 BCD Adder Circuit: Excess 3 Adder Circuit and Look-Ahead Adder Circuit

BCD binary numbers represent Decimal digits 0 to 9. A 4-bit BCD code is used to represent the ten numbers 0 to 9. Though the 4-bit Code allows 16 possibilities therefore, the first 10, 4-bit combinations are considered to be valid BCD combinations. Then, the latter six combinations are invalid and do not occur. BCD Code has applications in Decimal Number display Systems such as Counters and Digital Clocks.

BCD Numbers can be added together using BCD Addition. The BCD Addition is similar to normal Binary Addition except for the case when sum of two BCD digits exceeds 9 or a Carry is generated. During the Sum of two BCD numbers that exceeds 9 or a Carry is generated and 6 is added to convert it into a valid number from a invalid number.

The carry generated by adding a 6 to the invalid BDC digit is passed on to the next BCD digit. The addition of two BCD digits requires two 4-bit parallel adder circuits. One 4-bit parallel adder adds the two BCD digits. The BCD adder uses a circuit which checks the result at the output of the first adder circuit for determining whether the result has exceeded 9 or a carry has been generated.

When the circuit determines any of the two error conditions then, the circuit adds a 6 to the original result using the second adder circuit. The output of the second adder shall give the correct BCD output. If the circuit finds the result of the first Adder circuit to be a valid BCD number between 0 and 9 and no carry has been generated where the circuit adds a zero to the valid BCD result using the second adder. Thus, the output of the second adder also gives the same result.

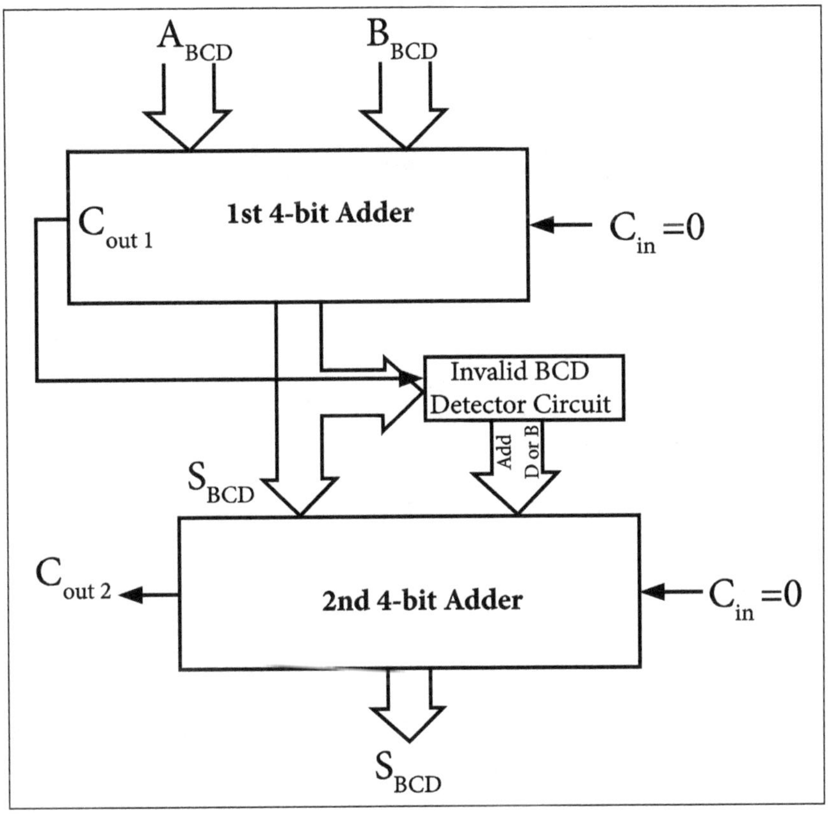

4-bit BCD adder.

3.4.1 Excess 3 Adder Circuit

To perform excess-3 addition we have to,

- Add two excess-3 numbers using binary addition.
- If carry = 1 → add 3 $(0\,0\,1\,1)_2$ to the sum of two digits = 0 subtract 3 $(0\,0\,1\,1)_2$ from the sum.

Example:

Let us perform the excess-3 addition of:

- 8, 6
- 1, 2

Solution:

a) 8 + 6

```
                      1 0 1 1  ──► Excess-3 for 8
                  +   1 0 0 1  ──► Excess-3 for 6
                      ───────
        Carry [1]     0 1 0 1
Add 3   0 0 1 1       0 0 1 1
        ───────       ───────
        [0 1 0 0]     [0 1 1 1] ──► Excess-3 for 14
            1             4
```

b) 1 + 2

```
                      0 1 0 0  ──► Excess-3 for 1
                  +   0 1 0 1  ──► Excess-3 for 2
                      ───────
           Carry [0]  1 0 0 1
Sub 3             -   0 0 1 1
                      ───────
                     [0 1 1 0] ──► Excess-3 for 3
                          3
```

3.4.2 Look-Ahead Adder Circuit

The carry look-ahead adder working can be understood better when we manipulate the Boolean expression dealing with the full adder.

The propagate P and generate G in a full-adder is given as:

$P_i = A_i \oplus B_i$ .....Carry propagate

$G_i = A_i . B_i$ ......Carry generate

It should be noted that both the propagate and generate signals will depend only on the input bits and it will be valid after one gate is delayed.

The output sum and the carryout for the new expressions are given by,

$S_i = P_i \oplus C_{i-1}$

$C_{i+1} = G_i + P_i C_i$

These equations will be generated in two cases which show that carry signal is as follows:

- When both bits A_i and B_i are 1.
- When either A_i or B_i is 1 and the carry-in C_i is 1.

Let us apply these equations for a 4-bit adder as follows:

- C1 = G0 + P0C0
- C2 = G1 + P1C1 = G1 + P1(G0 + P0C0) = G1 + P1G0 + P1P0C0
- C3 = G2 + P2C2 = G2 + P2G1 + P2P1G0 + P2P1P0C0
- C4 = G3 + P3C3 = G3 + P3G2 + P3P2G1 + P3P2P1G0 + P3P2P1P0C0

These expressions show that C2, C3 and C4 do not depend on its previous carry-in. Therefore C4 does not need to wait for C3 to propagate. As soon as the C0 is computed, C4 can reach steady state. The same is also true for C2 and C3. This is a two level Circuit.

The structure of Carry look-ahead adder can be divided into three parts as follows:

- Propagate/generate generator,
- Sum generator,
- Carry generator.

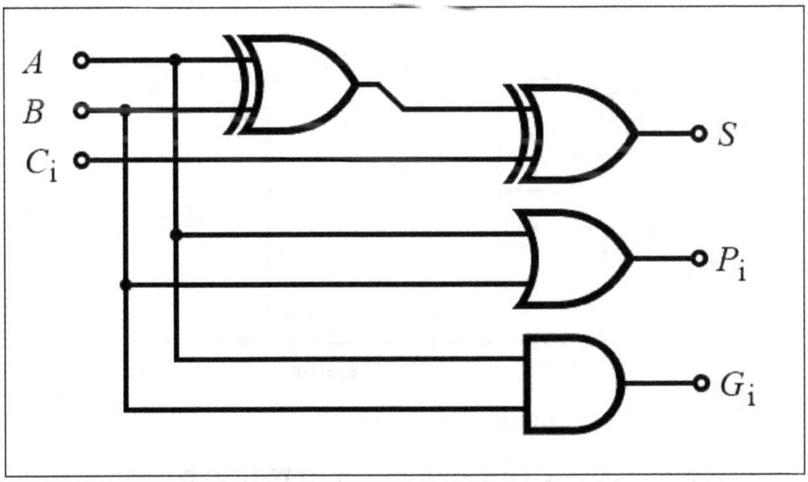

Propagate/ Generate generator and Sum generator.

The largest AND gate in the carry section will always have n+1 inputs and the number of AND gates requirements is n. Thus, it should have 16 AND gates where the biggest AND gate has 17 inputs for the design of 16 bit adder that needs the last carry generator section. The OR gate in this section also needs 17 inputs.

Thus, the size and fan-in of the gates needed to implement the Carry-Look-ahead adder is usually limited to four. Hence, a 4-bit Carry-Look ahead adder is designed as a block.

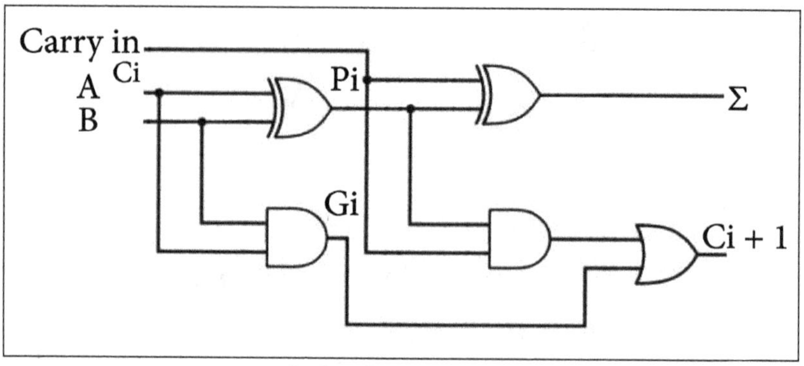

Look ahead carry generator.

The 4-bit Carry Look head adder block diagram is shown in below figure. The delay of such circuit is 4 levels of logic.

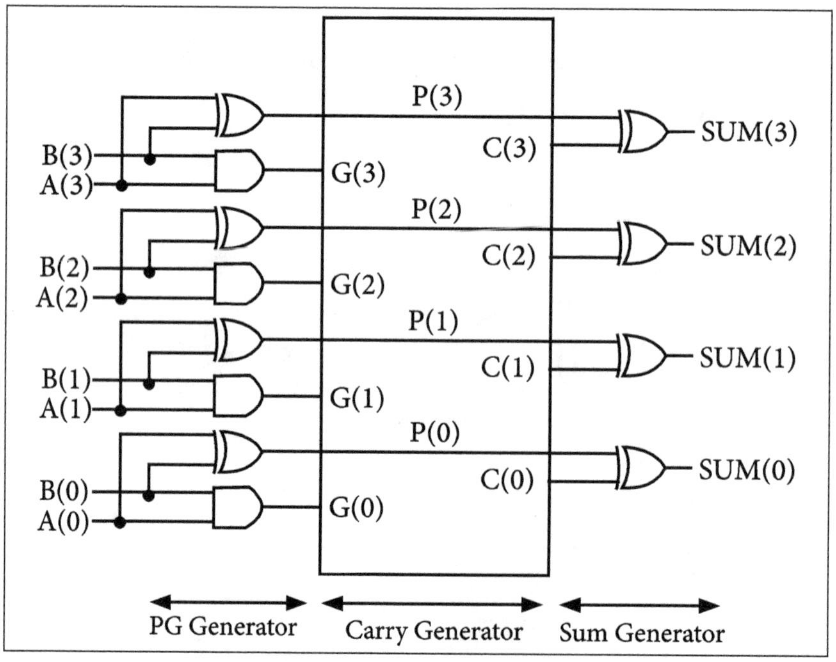

4-bit Carry look ahead adder.

But, in practice it is not possible to use the CLA to realize constant delay for the wider-bit adders though there will be a substantial loading capacitance and thus large delay and large power consumption.

The CLA has the fastest growing area and the power requirements with respect to the bit size.Speed also will drop with increase in bit size. Thus, even other techniques may be used.

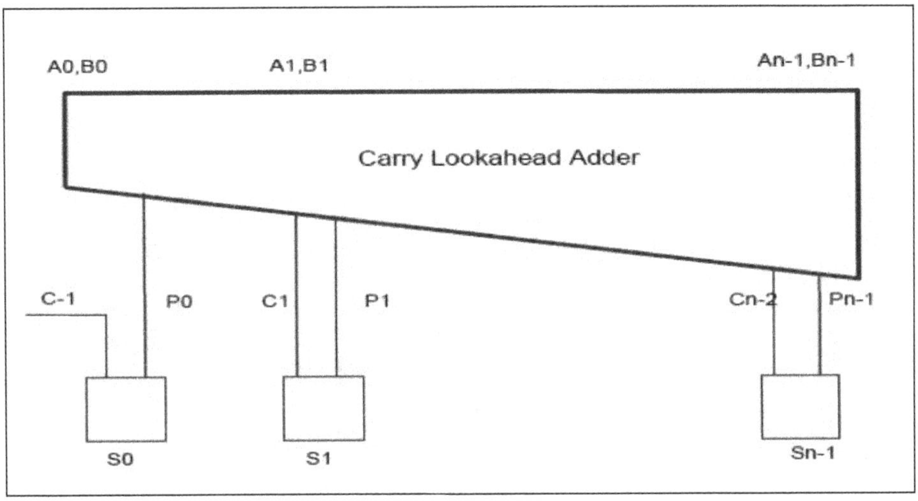

CLA Architecture.

For example, a 32-bit Carry-Look ahead adder is built by using 8 cascaded 4-bit Carry-Look ahead adders.

3.5 Design of Decoder and Realization of Boolean Functions

A decoder is a combinational circuit that converts binary information from n input lines to a maximum of 2^n output lines. A decoder is a multiple input multiple output logic circuit that converts coded inputs into coded outputs where the input and output codes are different.

$n - \text{to} - 2^{n\text{-Line Decoder}}$

Each AND gate has only n (in this case three) inputs and there are 2^n (in this case eight) outputs. Since there is no data input line to control, what used to be control inputs no longer serve that function. Instead, they are the data inputs to be decoded. This circuit is an example of an n-to-2^n-line decoder. Each output represents the minterm.

Output k is 1 whenever the combination of the input variable values is the binary equivalent of decimal k. Now, suppose that the data input line from the demultiplexer is not removed but retained and viewed as an enable input, the decoder now operates only when the enable is 1.

Viewed conversely, an n-to-2^n-line decoder with an enabled input can also be used as a demultiplexer, where the enable becomes the serial data input and the data inputs of the decoder become the control inputs of the demultiplexer.

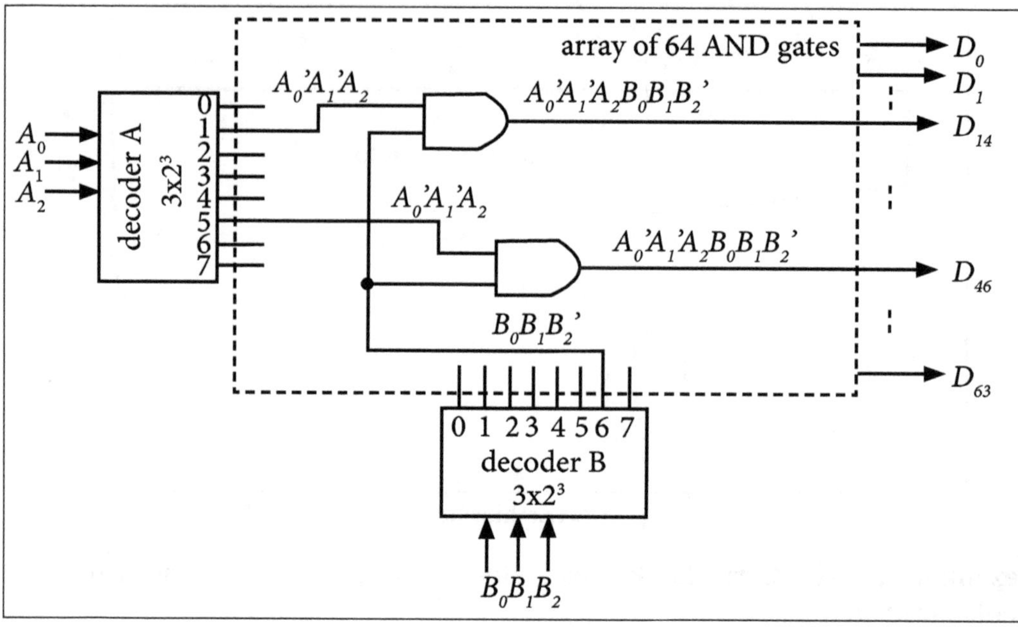

n-to-2n - Line Decoder.

Decoders are available as integrated circuits (MSI) whose n = 3 and n = 4 are quite common. However, there will be always some practical limitations on the fan-in decoders of higher order and are often designed using lower-order decoders interconnected with a network of other gates.

3 to 8 line Decoder

3 inputs are decoded into eight outputs. Each output represents one of the minterms of the 3 input variable.

EN	A	B	C	Y7	Y6	Y5	Y4	Y3	Y2	Y1	Yo
0	X	X	X	0	0	0	0	0	0	0	0
1	0	0	0	0	0	0	0	0	0	0	1
1	0	0	1	0	0	0	0	0	0	1	0
1	0	1	0	0	0	0	0	0	1	0	0
1	0	1	1	0	0	0	0	1	0	0	0
1	1	0	0	0	0	0	1	0	0	0	0
1	1	0	0	0	0	0	1	0	0	0	0
1	1	0	1	0	0	1	0	0	0	0	0
1	1	1	0	0	1	0	0	0	0	0	0
1	1	1	1	1	0	0	0	0	0	0	0

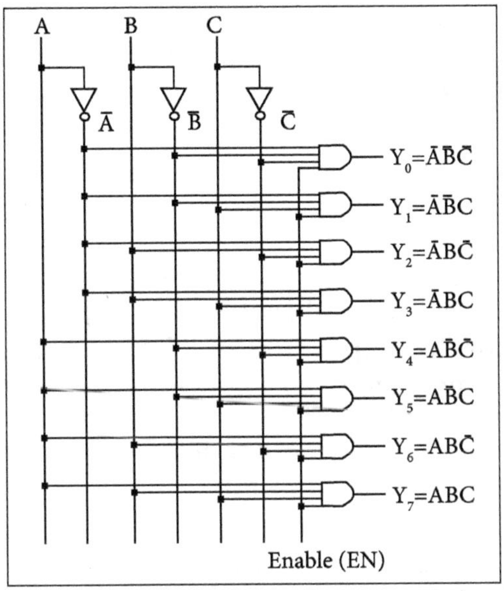

3 to 8 line decoder.

3.5.1 Demultiplexer

In addition to the data input there must be other inputs to control the transmission of the data to the appropriate data output line at any given time. This demultiplexer circuit having eight output lines is shown in the below figure (a). For the same number of control inputs, it has same number of AND gates. But now each AND gate output is a circuit output.

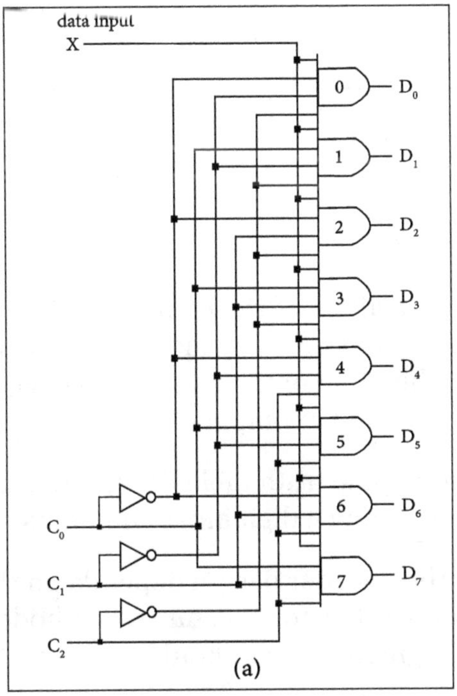

(a) Demux circuit.

Relatively, each gate having its own separate data input, single data line will form one of the inputs to each AND gate, the other AND inputs being control inputs. When the word formed by the control inputs $C_2C_1C_0$ which is the binary equivalent of decimal k, then the data input x is routed to output D_k. Otherwise, for a demultiplexer which has n control inputs, each AND gate output shall correspond to a minterm of n variables.

For a given combination of control inputs, only one minterm can take on the value 1, the data input is routed to the AND gate corresponding to this minterm. For example, if the logical expression for the output D_3 is $xC_2'C_1C_0$. Thus, when $C_2C_1C_0 = 011$, then $D_3 = x$ and all other D_i are equal to 0.

The complete truth table for the eight-output demultiplexer is given below:

Truth Table

Control inputs			Data outputs							
C_2	C_1	C_0	D_0	D_1	D_2	D_3	D_4	D_5	D_6	D_7
0	0	0	x	0	0	0	0	0	0	0
0	0	1	0	x	0	0	0	0	0	0
0	1	0	0	0	x	0	0	0	0	0
0	1	1	0	0	0	x	0	0	0	0
1	0	0	0	0	0	0	x	0	0	0
1	0	1	0	0	0	0	0	x	0	0
1	1	0	0	0	0	0	0	0	x	0
1	1	1	0	0	0	0	0	0	0	x

3.5.2 7-Segment Decoder

Step 1: The design involves analysis of the common cathode 7-segment display. The 7-segment display will consists of an arrangement of LEDs in an 'H' form. A truth table is constructed with the combination of inputs for each decimal number. Example, the decimal number 1 would command a combination of b and c.

Step 2: It involves constructing the truth table listing the 7 display input signals, decimal number and the corresponding 4 digit binary numbers.

Hence the truth table for the decoder design depends on the type of 7-segment display. As we mentioned above that for a common cathode seven-segment display, the output of decoder or segment driver should be active high in order to glow the segment.

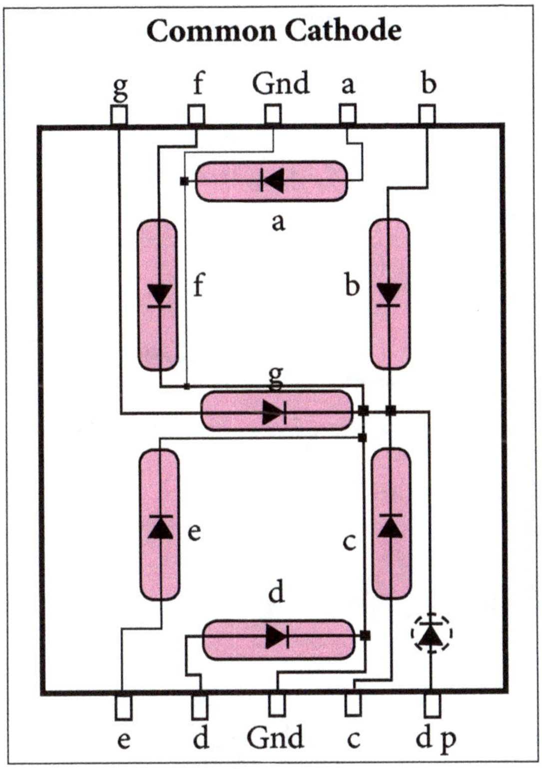

7 Segment LED.

The figure below shows the truth table of a BCD to seven-segment decoder with common cathode display. There are 7 different output columns corresponding to each of the 7 segments in truth table.

Take the column for segment 'a' shows the different combinations for which it is to be illuminated. Hence, 'a' is active for the digits 0, 2, 3, 5, 6, 7, 8 and 9.

Digit	A	B	C	D	a	b	c	d	e	f	g
0	0	0	0	0	0	0	0	0	0	0	1
1	0	0	0	1	1	0	0	1	1	1	1
2	0	0	1	0	0	0	1	0	0	1	0
3	0	0	1	1	0	0	0	0	1	1	0
4	0	1	0	0	1	0	0	1	1	0	0
5	0	1	0	1	0	1	0	0	1	0	0
6	0	1	1	0	0	1	0	0	0	0	0
7	0	1	1	1	0	0	0	1	1	1	1
8	1	0	0	0	0	0	0	0	0	0	0
9	1	0	0	1	0	0	0	0	1	0	0

From the above truth table, the equation obtained are as follows:

$$a = F1\,(A, B, C, D) = \Sigma m\,(0, 2, 3, 5, 7, 8, 9)$$
$$b = F2\,(A, B, C, D) = \Sigma m\,(0, 1, 2, 3, 4, 7, 8, 9)$$
$$c = F3\,(A, B, C, D) = \Sigma m\,(0, 1, 3, 4, 5, 6, 7, 8, 9)$$
$$d = F4\,(A, B, C, D) = \Sigma m\,(0, 2, 3, 5, 6, 8)$$
$$e = F5\,(A, B, C, D) = \Sigma m\,(0, 2, 6, 8)$$
$$f = F6\,(A, B, C, D) = \Sigma m\,(0, 4, 5, 6, 8, 9)$$
$$g = F7\,(A, B, C, D) = \Sigma m\,(2, 3, 4, 5, 6, 8, 9)$$

Thus, these are the Boolean expressions for each output functions.

Step 3: It involves constructing the K- map for each output term and then simplifying them for obtaining a logic combination of the inputs for each output.

K-Map Simplification

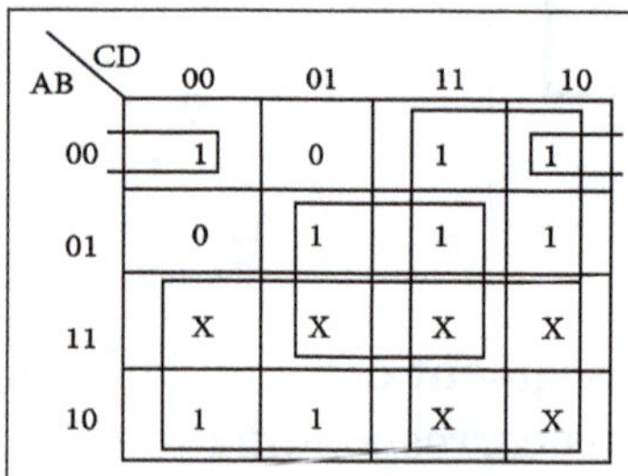

$a = A + C + BD + \bar{B}\bar{D}$

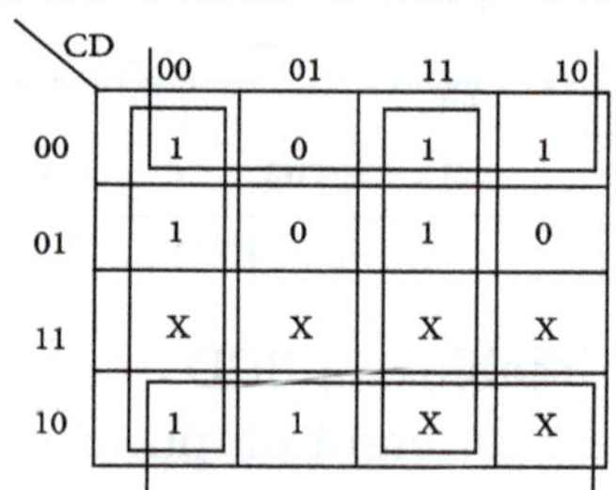

$b = \bar{B} + \bar{C}\bar{D} + CD$

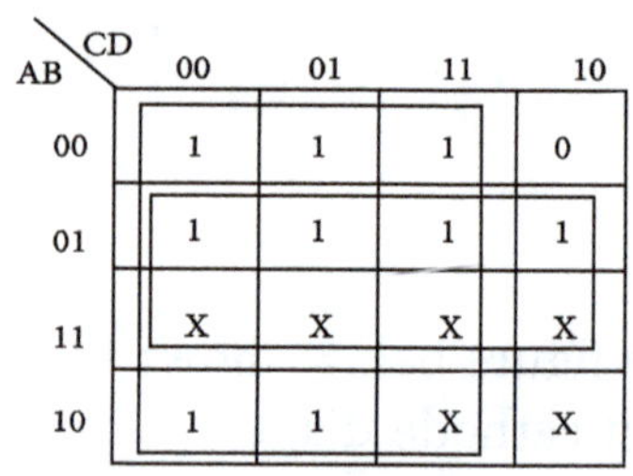

$c = B + \bar{C} + D$

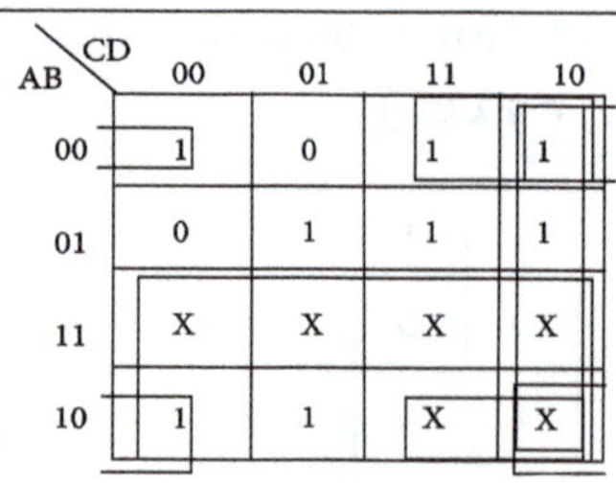

$d = \bar{B}\bar{D} + C\bar{D} + B\bar{C}D + \bar{B}C + A$

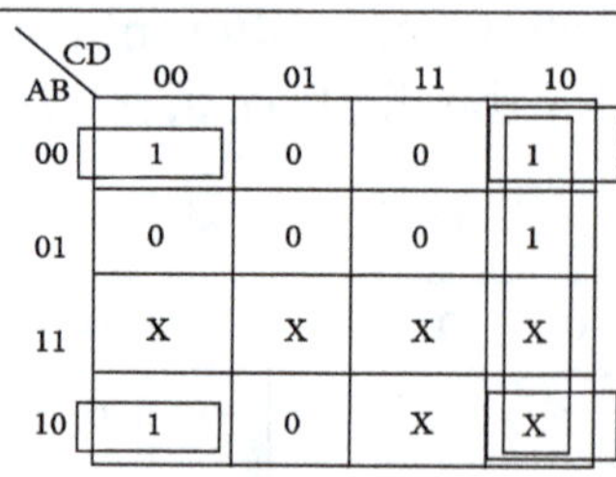

$e = \bar{B}\bar{D} + C\bar{D}$

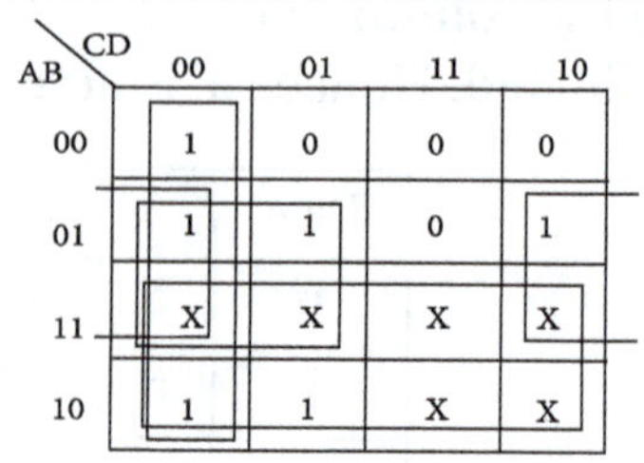

$f = A + \bar{C}\bar{D} + B\bar{C} + B\bar{D}$

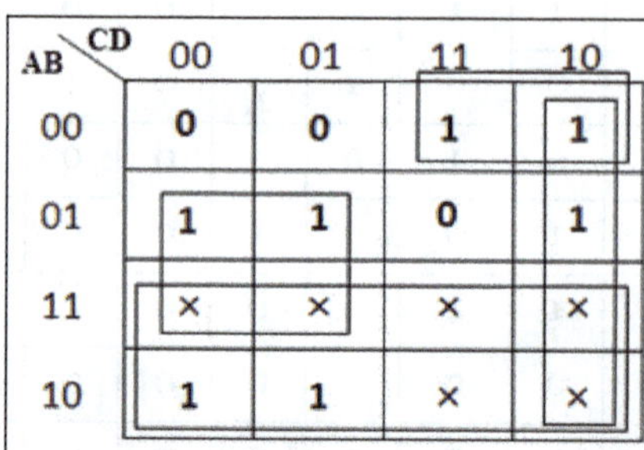

$g = \bar{B}C + C\bar{D} + B\bar{C} + B\bar{C} + A$

From the above simplification, it is given by:

$$a = A + C + BD + \overline{B}\overline{D}$$

$$b = \overline{B} + \overline{C}\overline{D} + CD$$

$$c = B + \overline{C} + D$$

$$d = \overline{B}\overline{D} + C\overline{D} + B\overline{C}D + \overline{B}C + A$$

$$e = \overline{B}\overline{D} + C\overline{D}$$

$$f = A + \overline{C}\overline{D} + B\overline{C} + B\overline{D}$$

$$g = A + B\overline{C} + \overline{B}C + C\overline{D}$$

We get the output values as shown above.

Step 4: Thus, the final step involves drawing a combinational logic circuit for each output signal. Once the task was accomplished, the combinational logic circuit can be drawn using 4 inputs (A,B,C,D)and a 7- segment display (a, b, c, d, e, f, g) as output.

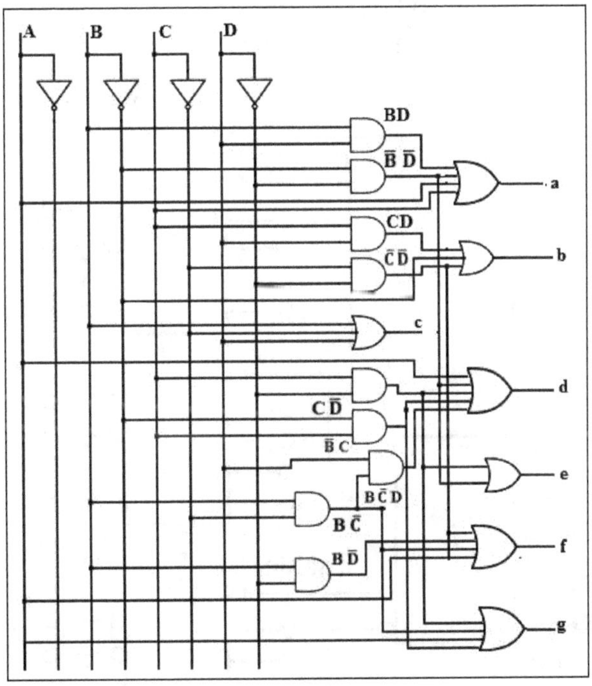

3.5.3 Higher Order de Multiplexing

Demultiplexer Tree

Since 4-line-to 16-line decoders are the largest available circuits in IC's, to meet the larger inputs need there should be a provision for expansion. This is made possible by using enable input terminal. The below figure (a) shows a 5-line-to 32-line decoder and figure(b) shows a 8-line-to 256-line decoder using 4-line-to 16-line decoders.

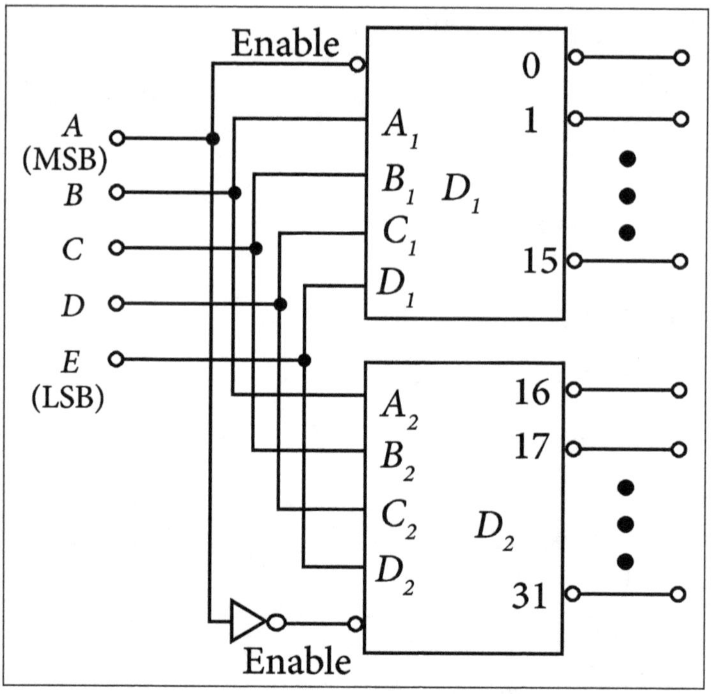

(a) 5-line-to 32-line decoder using two 4-line-to 16-line decoders.

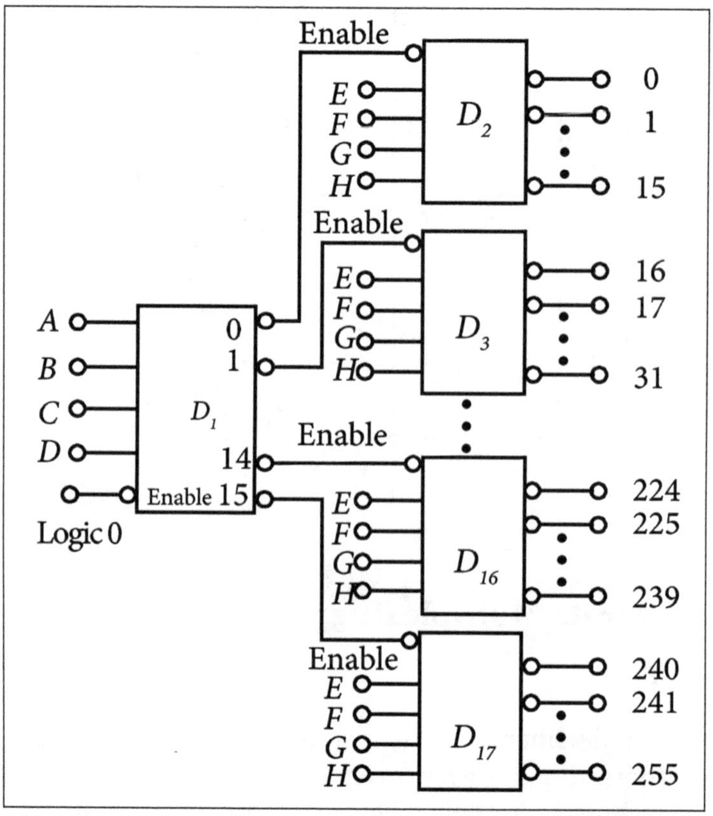

(b) 8-line-to 256-line decoder using two 4-line-to 16-line decoders.

Similarly, an m-line-to-n-line decoder can be implemented. However, if only a few codes of a large number need be recognized, the alternative approach, such as the one shown in the following figure (c), can be used. This is connected to detect the digital number 000111.

The most significant 4-bits are applied at ABCD inputs and the least-significant bits are applied at EFGH inputs. The output goes low when the most-significant bits are 0001 and the least-significant bits are 1111. The circuit can be expanded to detect other 8-bit codes.

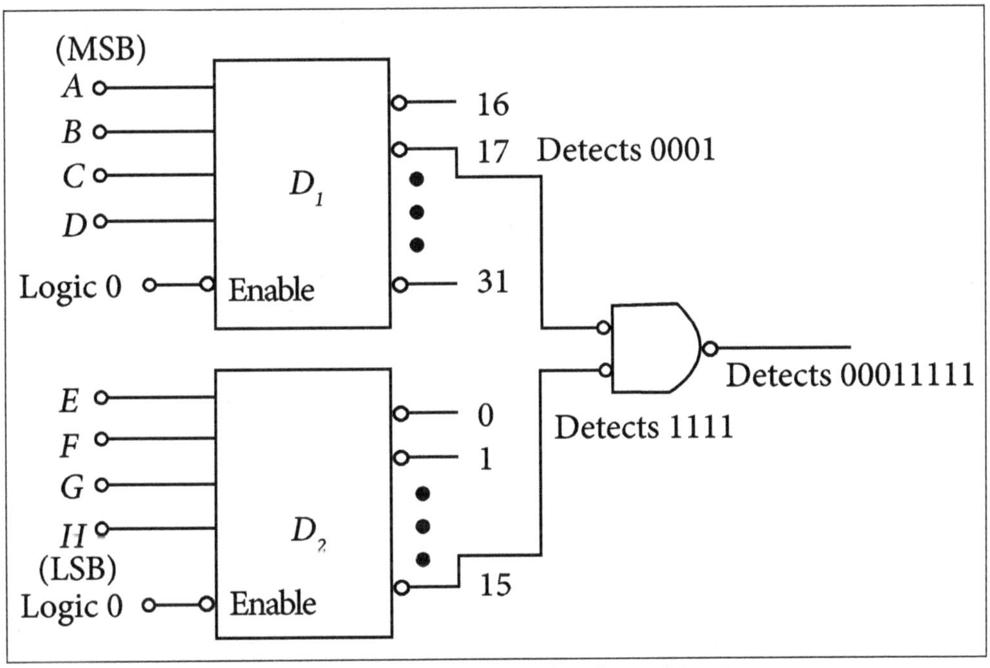

(c) An alternative approach to decode combinations.

3.5.4 Encoder

An encoder is the digital circuit that performs the inverse operation of a decoder. An encoder has 2^n input lines and n output lines. In the encoder, the output lines generate the binary code corresponding to the input value. An encoder is a combinational circuit that performs the inverse operation of a decoder.

When the device output code has fewer bits than the input code has, the device is usually called an encoder. E.g., 2^n-to-n priority encoders. It is a 2^n-to-n binary encoder, where it has only 2^n inputs and the output is the n-bit binary number corresponding to the active input.

The figure below shows the general structure of the encoder circuit. As shown in the figure below, the decoded information is presented as 2^n inputs producing n possible outputs.

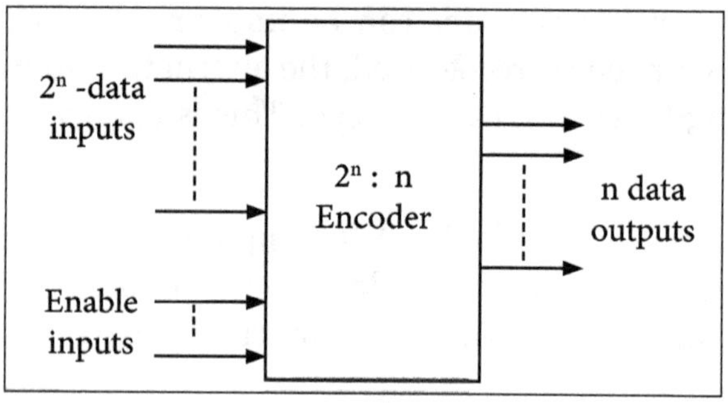

Encoder

3.5.5 Multiplexer, Higher Order Multiplexing

Multiplexer is a digital switch. It allows digital information from several sources to be routed onto a single output line. Multiplexer is a special type of combinational circuit. It has n data inputs, one output and select inputs with $2^m = n$.

Depending on the digital code applied at the select inputs, one out of a data sources is selected and transmitted to the single output Y. E is called as a strobe or enable input which is useful for cascading. It is generally an active low terminal that means it will perform required operation when it is low.

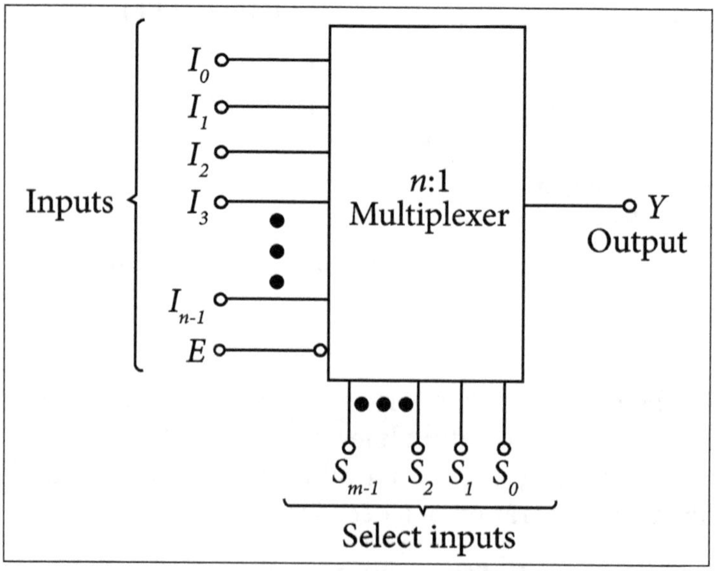

Multiplexer

Applications of Multiplexer

It is used:

- In the implementation of combinational circuits.
- As function generators.

Data Routing

Multiplexers can be used to route the data from one of several sources to one destination.

Control Sequences

A Multiplexer can also be used as a part of control sequences.

2 x 2 Multiplexer

Multiplication of binary number is performed in the same way as in decimal numbers. The multiplicand is multiplied by each bit of the multiplier which starts from the least significant bit. Each such multiplication forms a partial product. Successive partial products are shifted one position to the left. The final product is obtained from the sum of partial products.

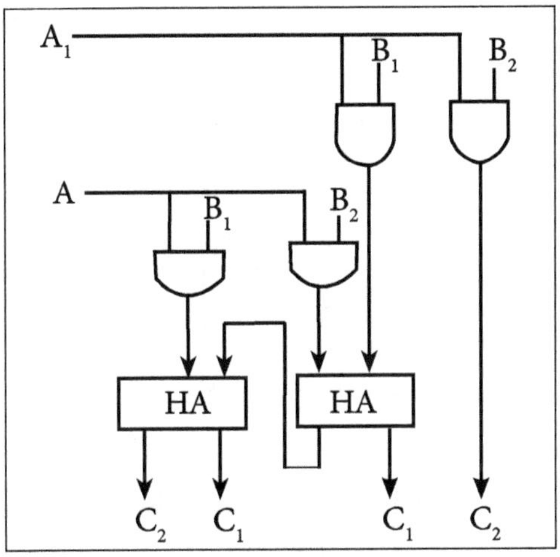

2 x 2 multiplexer.

The multiplication of 2 bit number is considered. The multiplicand bits are B_1 and B_0, the multiplier bits are A_1 and A_0 and the product is $C_3C_2C_1C_0$. The first partial product is formed by multiplying A_0 by B_1B_0.

The multiplication of two bits such as A_0 and B_0 produces 1 if both bits are 1, otherwise produces 0. This is identical to AND operation. Therefore, the partial product can be implemented with AND gates as,

$$\begin{array}{cccc} & & B_1 & B_0 \\ & & A_1 & A_0 \\ \hline & & A_0B_1 & A_0B_0 \end{array}$$

$$\begin{array}{cccc} & A_1B_1 & A_1B_0 & \\ \hline C_3 & C_2 & C_1 & C_0 \end{array}$$

The second partial product is formed by multiplying A_1 by B_1B_0 and shifted one position to left. The two partial products are added with two half adder circuits. The LSB of the product does not have to go through an adder since it is formed by the output of the first AND gate.

Higher Order Multiplexing

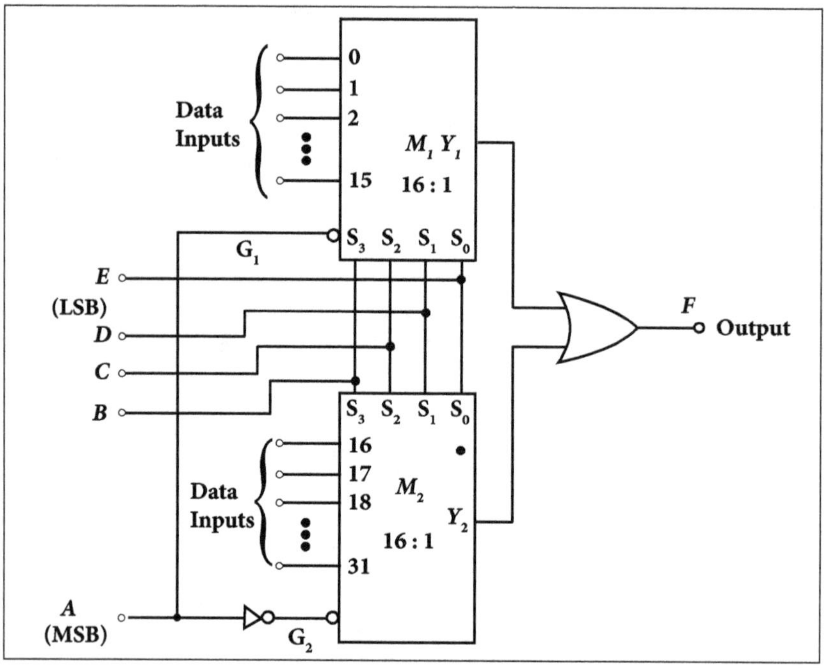

(a) 16 : 1 mux with OR gate.

The 16-to-1 multiplexers are the largest available IC's. Hence, to meet the larger input needs there must be a provision for expansion. This can be achieved with the help of enable or strobe inputs and multiplexer stacks or trees are designed. Two commonly used methods for this purpose are illustrated in the above figure (a) and (b).

The circuit of the above figure (a) uses two 16:1 multiplexers (M_1 and M_2) for the realization of a 32:1 multiplexer. The lower order of 16 data input lines $(I_0 - I_{15})$ are applied at the data input terminals of the multiplexer M_1 and the higher order 16 data input lines $(I_{15} - I_{31})$ are applied at the data input terminals of the multiplexer M_2.

For a 32:1 multiplexer, the number of select input lines is required to be 5. The 5-bit select inputs are ABCDE. The most significant select input bit A is applied at G_1 and $\bar{A}$ is applied at G_2. B, C, D and E are connected to S_3, S_2, S_1 and S_0 inputs of both the multiplexers. The output F of the multiplexer is obtained by using an OR gate, where $F = Y_1 + Y_2$.

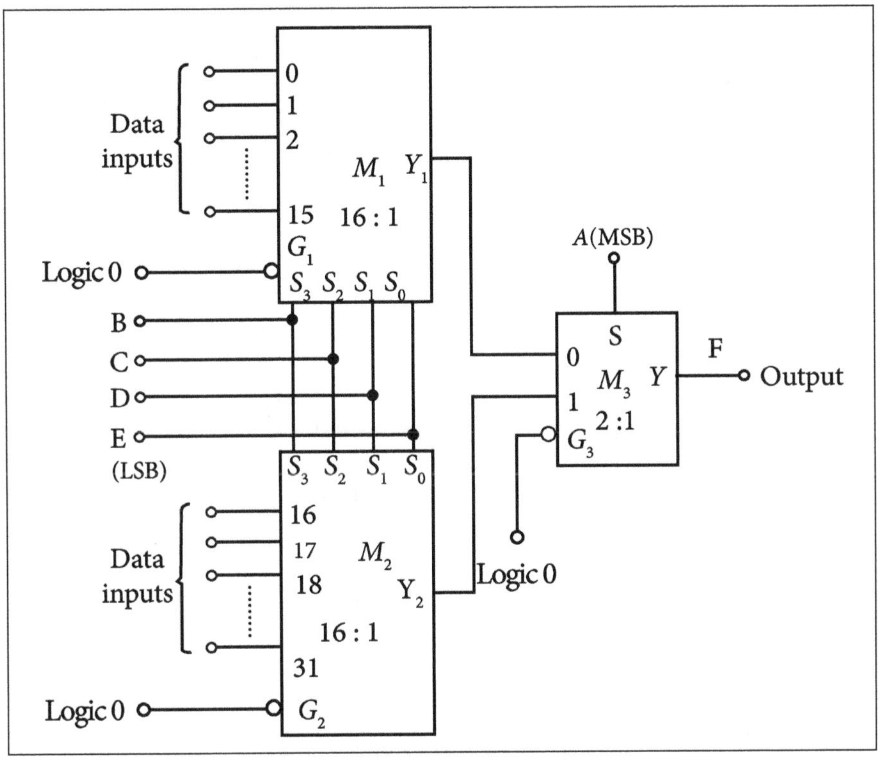

(b) 16 : 1 mux with 2 :1 mux.

When A = 0, the multiplexer M_1 is enabled and M_2 is disabled, thereby allowing one of the lower order 16 bits to Y_1, depending upon the value of BCDE. $Y_2 = 0$. Therefore, $F = Y_1$. Similarly, when A = 1, M_2 is enabled and M_1 is disabled, then $F = Y_2$. The above figure (b) uses a 2:1 multiplexer instead of an OR gate.

When A = 0, $F = Y_1$ and when A = 1, $F = Y_2$. Thus, both the circuits perform similar operation. These two general techniques can be used to expand to an n input multiplexer without any difficulty.

3.5.6 Realization of Boolean Functions Using Decoders and Multiplexers

Let us consider the following truth table that describes a function of 4 Boolean variables.

A 16 to 1 Multiplexer with A, B, C, and D applied to its S_3, S_2, S_1, and S_0 inputs respectively would select one of its 16 inputs for each of the 16 possible combinations of A, B, C, and D. We can implement the function described by the truth table by connecting a voltage source for logic level 1 or ground for a logic level 0 to each of the Multiplexer inputs corresponding to the required value of the function associated with the combination of A, B, C, and D that selected the input. Therefore, the inputs to the Multiplexer will be the same as the F entries in the truth table provided A, B, C, and D are connected to the Multiplexer select inputs in the right order.

A	B	C	D	F
0	0	0	0	0
0	0	0	1	0
0	0	1	0	1
0	0	1	1	1
0	1	0	0	0
0	1	0	1	1
0	1	1	0	0
0	1	1	1	1
1	0	0	0	0
1	0	0	1	0
1	0	1	0	1
1	0	1	1	1
1	1	0	0	0
1	1	0	1	0
1	1	1	0	0
1	1	1	1	1

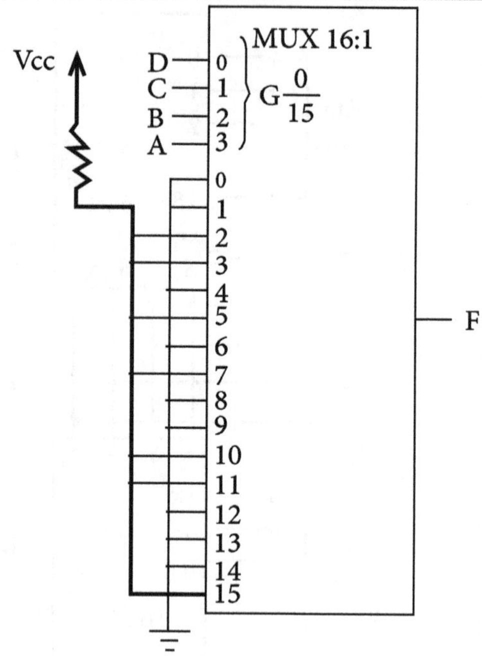

(a) 16 to 1 Multiplexer implementation.

Let us consider instead of using a 16 to 1 Multiplexer, we use an 8 to 1 Mux. If we choose to connect A, B, and C to the inputs of the Multiplexer, then for each combination of A, B and C although only one Mux input is selected, we need to realize two potentially different values of F. If they are different, they will be functions of D (either D or $\bar{D}$). To see this, we partition the truth table into sections where A, B and C have the same value. This is shown in the below figure (b).

A	B	C	D	F	
0	0	0	0	0	0
0	0	0	1	0	
0	0	1	0	1	1
0	0	1	1	1	
0	1	0	0	0	D
0	1	0	1	1	
0	1	1	0	0	D
0	1	1	1	1	
1	0	0	0	0	0
1	0	0	1	0	
1	0	1	0	1	1
1	0	1	1	1	
1	1	0	0	0	0
1	1	0	1	0	
1	1	1	0	0	D
1	1	1	1	1	

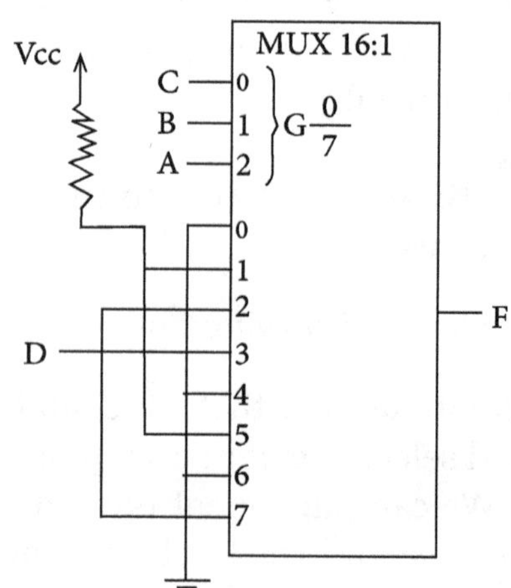

(b) 8 to 1 Multiplexer implementation.

This idea can be extended to a 4 to 1 Mux implementation. In this case we partition the table into sections for which A and B has the same value. This is shown in the below figure (c). Now Multiplexer inputs will be either 1, 0 or some function of C and D. K-maps may be used to simplify the functions of C and D for each of the table partitions, but in the case of two variables, the functions are usually obvious.

A	B	C	D	F	
0	0	0	0	0	
0	0	0	1	0	*C*
0	0	1	0	1	
0	0	1	1	1	
0	1	0	0	0	
0	1	0	1	1	*D*
0	1	1	0	0	
0	1	1	1	1	
1	0	0	0	0	
1	0	0	1	0	*C*
1	0	1	0	1	
1	0	1	1	1	
1	1	0	0	0	
1	1	0	1	0	*CD*
1	1	1	0	0	
1	1	1	1	1	

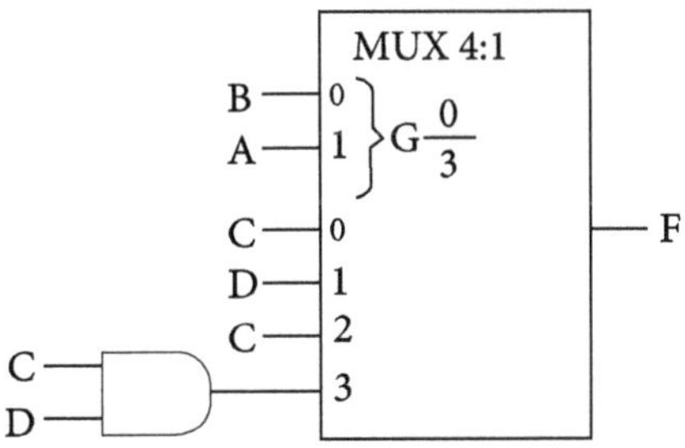

(c) 4 to 1 Multiplexer implementation.

The final possibility, shown in the following figure (d), is a 2 to 1 Mux implementation. In this case only A is used as a select input and the table is partitioned into two parts based on A. Inputs to the Mux will be functions of B, C and D. Here it is more likely that you will need to use K-maps to simplify the functions.

A	B	C	D	F	
0	0	0	0	0	
0	0	0	1	0	
0	0	1	0	1	$\bar{B}C+BD$
0	0	1	1	1	
0	1	0	0	0	
0	1	0	1	1	
0	1	1	0	0	
0	1	1	1	1	
1	0	0	0	0	
1	0	0	1	0	
1	0	1	0	1	$\bar{B}C+CD$
1	0	1	1	1	
1	1	0	0	0	
1	1	0	1	0	
1	1	1	0	0	
1	1	1	1	1	

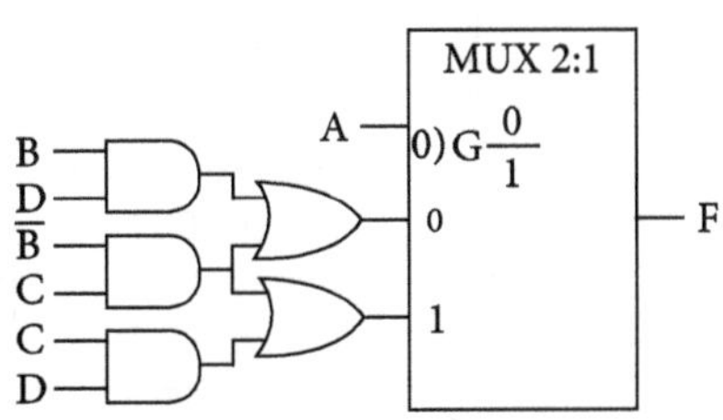

(d) 2 to 1 Multiplexer implementation.

It should be apparent that there is no explicit importance to the order of A, B, C and D in the truth table. If we rearranged them, we would have chosen different variables as Mux select inputs, and we might end up using less logic in the rest of the circuit. Let we assume that we are rearranging the table so that B is the Mux select input.

Problems

1. Let us see the implementation of Boolean function using 8:1 multiplexer.

$$F(A,B,C,D) = \bar{A}B\bar{D} + ACD + \bar{B}CD + \bar{A}\bar{C}D$$

Solution:

Given:

$$F(A,B,C,D) = \bar{A}B\bar{D} + ACD + \bar{B}CD + \bar{A}\bar{C}D$$

The given Boolean expression is not in standard SOP form. The conversion to SOP is:

$$\begin{aligned} F(A,B,C,D) &= \bar{A}B\bar{D} + ACD + (\bar{B} + \bar{B}) + \bar{B}CD(A + \bar{A}) \\ &\quad + \bar{A}\bar{C}D(B + \bar{B}) \\ &= \bar{A}BCD + \bar{A}B\bar{C}\bar{D} + ABCD + A\bar{B}CD + A\bar{B}CD \\ &\quad + \bar{A}\bar{B}CD + \bar{A}B\bar{C}D + \bar{A}\bar{B}\bar{C}D \\ &= \bar{A}BC\bar{D} + \bar{A}B\bar{C}\bar{D} + ABCD + A\bar{B}CD \\ &\quad + \bar{A}\bar{B}CD + \bar{A}B\bar{C}D + \bar{A}\bar{B}\bar{C}D \end{aligned}$$

The Truth Table for this standard SOP is given below:

No.	Minterms	A	B	C	D	Y
0		0	0	0	0	0
1	$\bar{A}\bar{B}\bar{C}D$	0	0	0	1	1
2		0	0	1	0	0
3	$\bar{A}\bar{B}CD$	0	0	1	1	1
4	$\bar{A}B\bar{C}\bar{D}$	0	1	0	0	1
5	$\bar{A}B\bar{C}D$	0	1	0	1	1
6	$\bar{A}BC\bar{D}$	0	1	1	0	1
7		0	1	1	1	0
8		1	0	0	0	0
9		1	0	0	1	0

10		1	0	1	0	0
11	$A\bar{B}CD$	1	0	1	1	1
12		1	1	0	0	0
13	-	1	1	0	1	0
14		1	1	1	0	0
15	ABCD	1	1	1	1	1

Implementation Table

	D_0	D_1	D_2	D_3	D_4	d_5	d_6	d_7
$\bar{A}$	0	1	2	3	4	5	6	7
A	8	9	10	11	12	13	14	15
	0	$\bar{A}$	0	1	$\bar{A}$	$\bar{A}$	$\bar{A}$	A

MUX Implementation

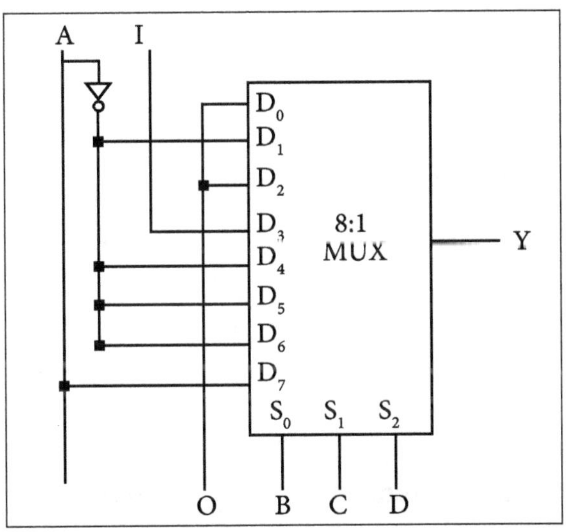

2. Let us implement the following with a multiplexer.

$$F(A, B, C) = \Sigma(1, 2, 4, 5)$$

Solution:

Given:

$$F(A, B, C) = \Sigma(1, 2, 4, 5)$$

Two of the variable B and C, are applied to the selection lines S_1 and S_0 respectively.

The truth table of multiplexer is given as:

Minterm	A	B	C	F
0	0	0	0	0
1	0	0	1	1
2	0	1	0	1
3	0	1	1	0
4	1	0	0	1
5	1	0	1	1
6	1	1	0	0
7	1	1	1	0

The Implementation Table

	D_0	D_1	D_2	D_3
$\bar{A}$	0	①	②	3
A	④	⑤	6	7
	A	1	$\bar{A}$	0

The multiplexer implementation is given as,

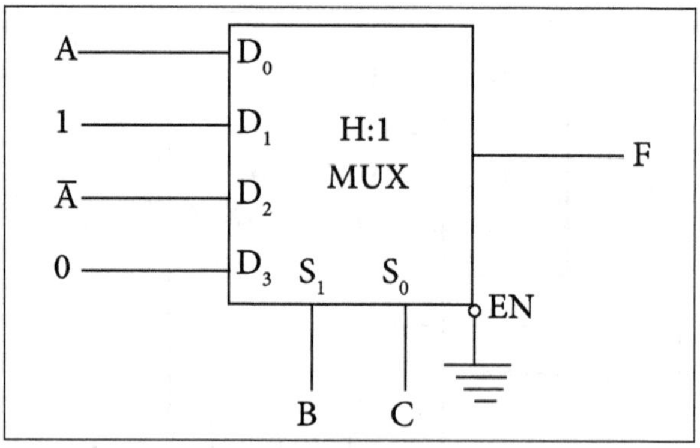

3. Let us implement the following function f(A, B, C) = Σm(0, 3, 5) using 4: 1 multiplexer.

Solution:

Given:

$$f(A, B, C) = \Sigma m(0, 3, 5)$$

Variables n = 3 (A, B, C)

Select lines = n – 1 = 2(S_0, S_1)

2^{n-1} to 1 MUX= 4: 1 MUX

Input lines $= 2^{n-1} = 4(I_0, I_1 I_2, I_3)$

Implementation Table

	I_0	I_1	I_2	I_3
$\hat{A}$		1	2	
A	4		6	7
	$\bar{A}$	A	0	$\bar{A}$

$I_0 = \bar{A}, I_1 = A, I_2 = 0, I_3 = \bar{A}$

Multiplexer Implementation

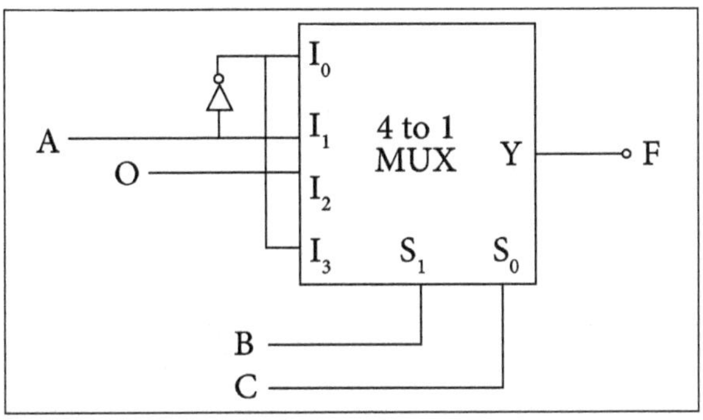

3.5.7 Priority Encoder

This is a special type of encoder. Priorities are given to the input lines. If two or more inputs lines are '1' at the same time, then the input lines with highest priority will be considered. A priority encoder is the practical form of an encoder. The encoders that are available in IC form are all priority encoders.

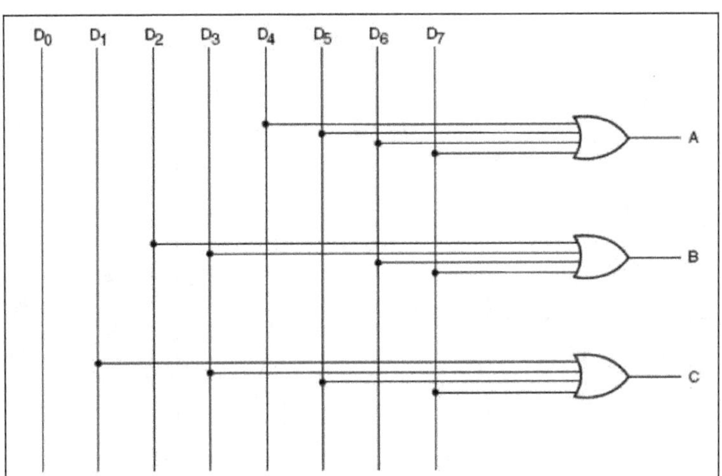

Priority Encoder.

In this type of encoder, a priority is assigned to each of the input so that when more than one input is simultaneously active, the input with the highest priority is encoded. Let us also assume that the input lines D_2, D_4 and D_7 are all simultaneously in logic 1' state. In such cases, only D_7 will be encoded and its output will be 111.

3.6 4-Bit Digital Comparator

It is a combinational circuit that compares two numbers and determines their relative magnitude. The output of comparator is usually 3 binary variables indicating:

A > B

A = B

A < B

Comparators

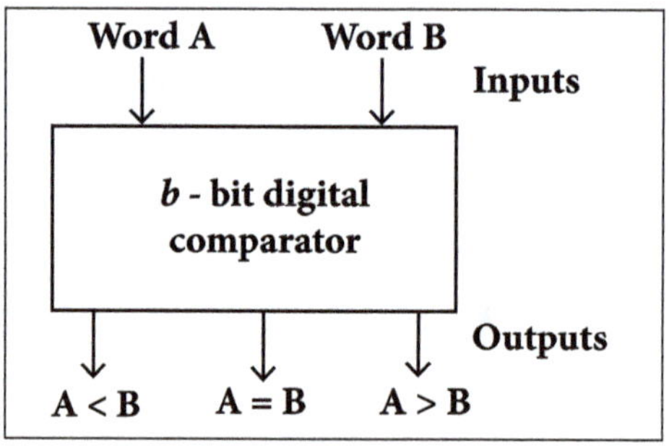

Comparator

Digital comparator is a combinational circuit, designed to compare the two n-bit binary words applied at its input. Comparator has three outputs A > B, A = B and A < B depending on the result of comparison of one of their outputs.

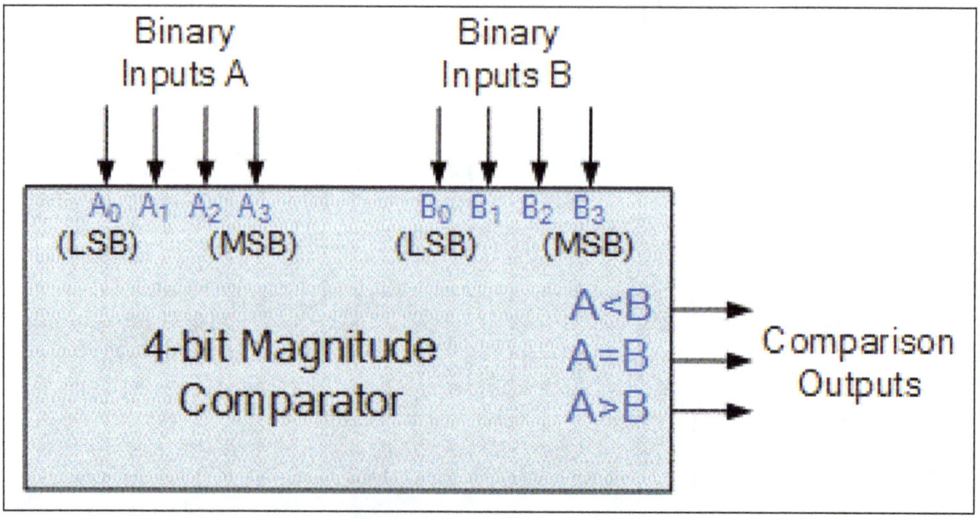

4-Bit Digital Comparator.

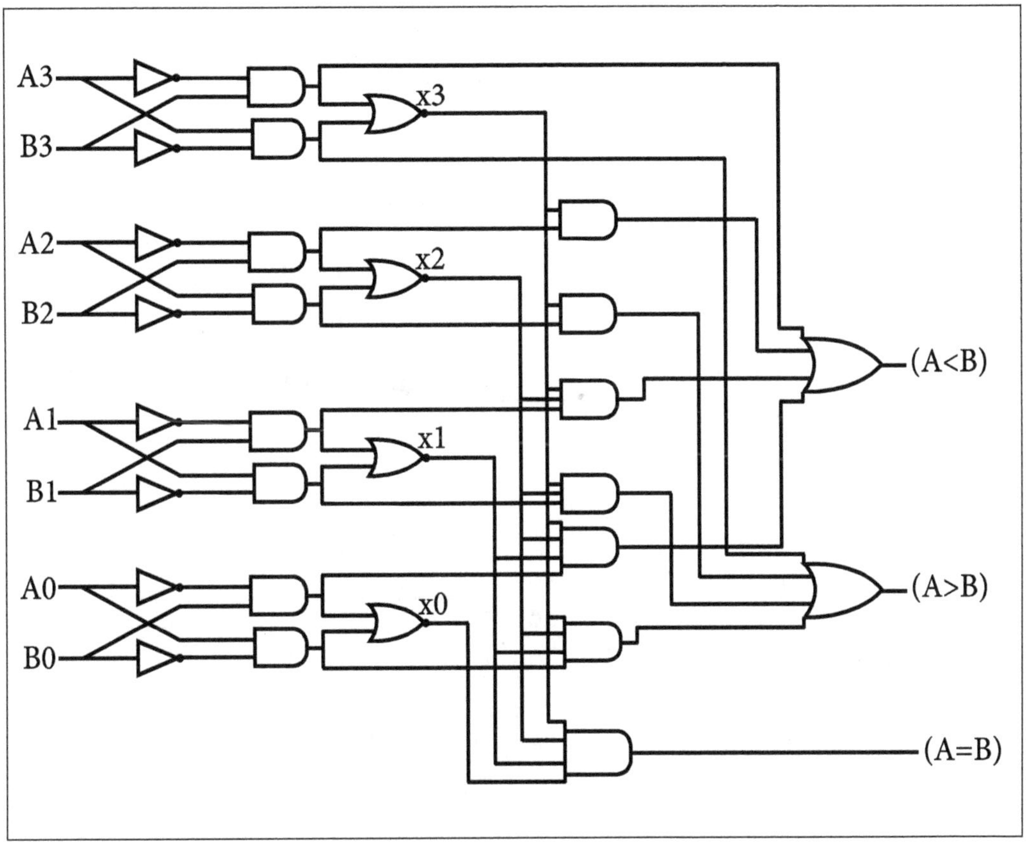

4-bit magnitude comparator.

The method for binary numbers with more than 2 bits can also be found in the similar way. The above figure shows the 4-bit magnitude comparator.

Input A = A3A2A1A0

B = B3B2B1B0

From the above diagram the input are as follows,

1. A =h B : A3 = B3, A2 = B2, A1 = B1, A0 = B0

$$xi = AiBi + Ai'Bi'$$

$$XOR-Invert = (AiBi'+Ai'Bi)' = (Ai' + Bi)(Ai + Bi')$$

$$= Ai'Ai + Ai'Bi' + AiBi + BiBi'$$

$$= AiBi + Ai'Bi'$$

Output: x3x2x1x0

2. A > B

Output: A3B'3 + x3A2B'2 + x3x2A1B'1+ x3x2x1A0B'0

3. (A < B)

Output: A'3B3 + x3A'2B2 + x3x2A'1B1+ x3x2x1A'0B0

Truth Table

Comparing Inputs				Outputs		
A3,B3	A2,B2	A1,B1	A0,B0	A > B	A < B	A = B
A3 > B3	X	X	X	H	L	L
A3 < B3	X	X	X	L	H	L
A3 = B3	A2 > B2	X	X	H	L	L
A3 = B3	A2 < B2	X	X	L	H	L
A3 = B3	A2 = B2	A1 > B1	X	H	L	L
A3 = B3	A2 = B2	A1 < B1	X	L	H	L
A3 = B3	A2 = B2	A1 = B1	A0 > B0	H	L	L
A3 = B3	A2 = B2	A1 = B1	A0 < B0	L	H	L
A3 = B3	A2 = B2	A1 = B1	A0 = B0	H	L	L
A3 = B3	A2 = B2	A1 = B1	A0 = B0	L	H	L
A3 = B3	A2 = B2	A1 = B1	A0 = B0	L	L	H
H = High Voltage Level; L = Low Voltage Level; X = Don't Care						

4

Introduction of PLD's

4.1 PROM, PAL and PLA-Basics Structures

ROM: Read-Only Memory

Information cannot be easily stored. The bits are initially decoded and are never changed thereafter.

ROM Organization

It is a read only memory.

We cannot write data in this memory. It is non-volatile memory i.e., it can hold data even if power is turned off.

There are four types of ROM. They are Masked ROM, PROM, EPROM and EEPROM or E^2PROM.

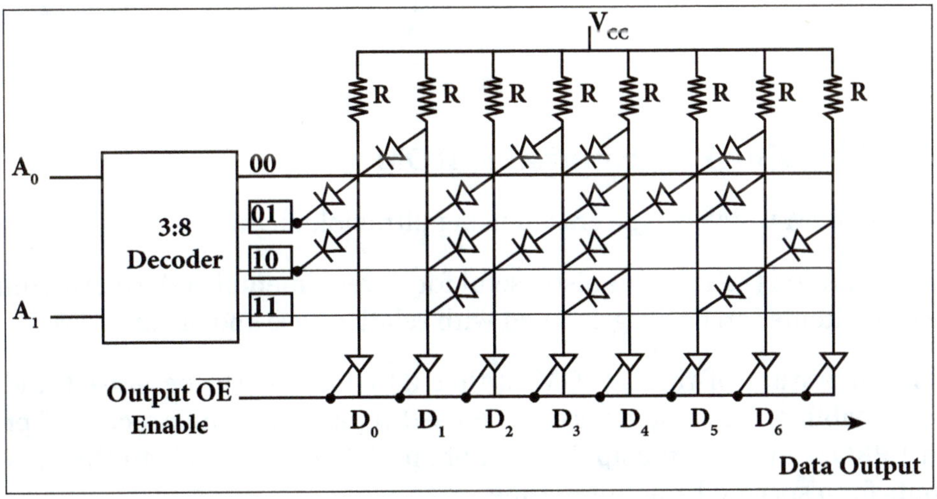

ROM

PROM - Programmable ROM

Bits can be set on a programming bench by burning fusible links or equivalent. This technology is also used for programmable array logic (PALs).

The figure given below shows four byte PROM. It has diodes in every bit position therefore, the output is initially all 0's.

Each diode however has a fusible link in series with it. By addressing bit and applying proper current pulse at the corresponding output, we can blow out the fuse, storing logic 1 at that bit position.

The PROM programmer selectively burns the fuses according to the bit pattern to be stored. This process is known as burning of PROM. It is one time programmable. Once programmed, the information stored is permanent.

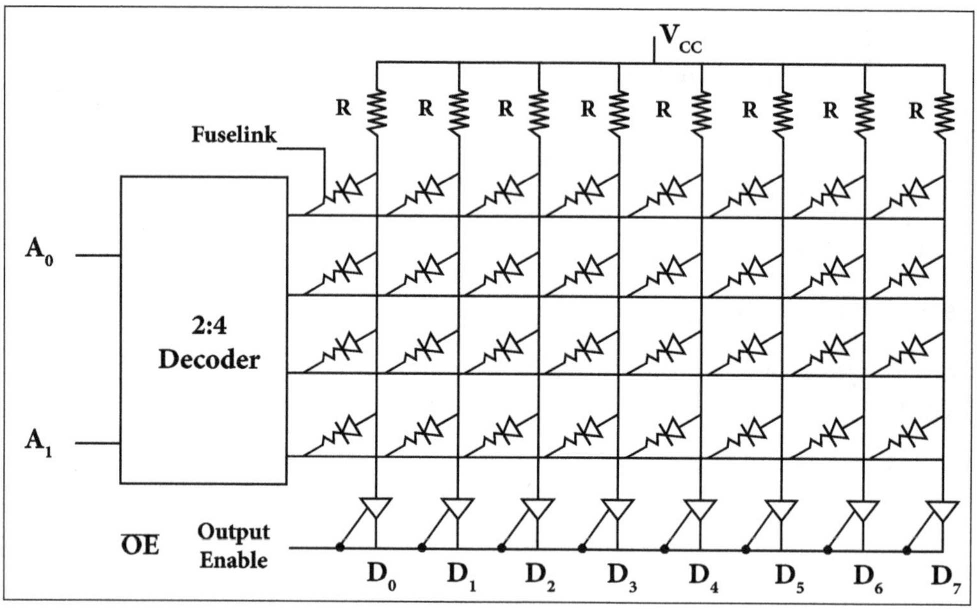

PROM

EPROM - Erasable Programmable ROM

It is a type of ROM which can be erased using ultraviolet light.

The erasable PROM (EPROM) is also useful for developmental and experimental work because its contents can be programmed with relative ease and at small cost.

The principal feature of the EPROM is that its contents can be erased and reprogrammed, enabling the same device to be used repeatedly. The process of programming an EPROM involves the application of special voltage levels to the appropriate chip inputs for a specified amount of time.

The complete programming process can take up several minutes for one EPROM chip. To erase an EPROM, the cells must be exposed to relatively intense UV radiation and it has transparent quartz window. This erasing process typically requires 15 to 30 minutes of exposure to UV rays.

Once erased, the EPROM can be reprogrammed. Once an EPROM has been programmed an opaque cable is placed over the window to prevent stray radiation from causing gradual erasure.

The EPROMs are available in a wide range of capacities and access times. A popular line of commercially available EPROMs is the 27xx series.

Difference between PROM and EPROM

PROM	EPROM
PROM is Programmable Read Only Memory. It consists of a set of fixed AND gates connected to a decoder and a programmable OR array. The programmable OR gates implement the Boolean functions in sum of minterms.	EPROM is Erasable and Programmable Read Only Memory. The ROM can be erased and programmed again and again.

Electrically Alterable or Erasable Prom (Eaprom Or Eeprom)

The EAPROM or EEPROM can be programmed (modified) or erased while it is still in the circuit and used like a nonvolatile read/write memory.

EAPROM/EEPROM uses an electric charge to erase the contents.

Some types of EAPROM/EEPROM are more versatile. Individual cells can be reprogrammed by reversing the voltage used to create a zero.

There are some timing constraints that causes more time for erasure or programming than to read data from the part. Some EAPROM/EEPROM has a word or byte erase mode.

Programmable Array Logic (PAL)

Programmable Array Logic (PAL) is a type of Programmable Logic Device (PLD) used to realize a particular logical function. PALs comprise of an AND gate array followed by an OR gate array as shown in the below figure.

However, it is to be noted that here only the AND gate array is programmable unlike the OR gate array which has a fixed logic. This is because the inputs are fed to the AND gates through fuses, which act as programmable links.

PAL is a family of programmable logic device semiconductors used to implement logic functions in digital circuits introduced by Monolithic Memories, Inc. (MMI) in March 1978. MMI obtained a registered trademark on the term PAL for use in "Programmable Semiconductor Logic Circuits". The trademark is currently held by Lattice Semiconductor.

Programmable-AND and fixed-OR structure of PAL's make them less flexible from programming point of view when compared with Programmable Logic Arrays (PLA's). However, due to the same reason PAL's are less expensive than PLA's.

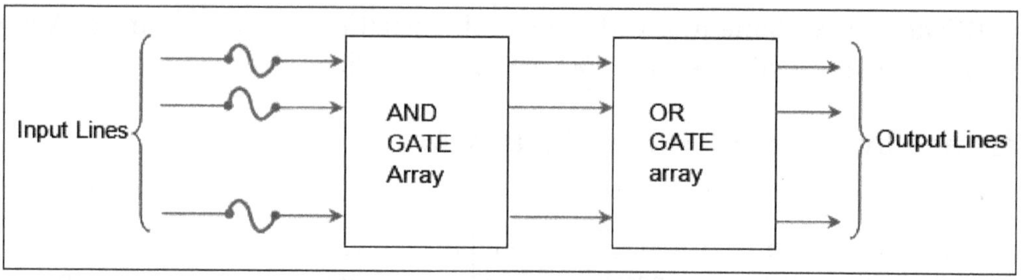

Programmable Array Logic (PAL).

The below figure shows the internal structure of a PAL with m inputs and n outputs. Each of the input line is showed to pass through the buffers and/or inverters. All of these inputs are connected to each and every AND gate present in the PAL.

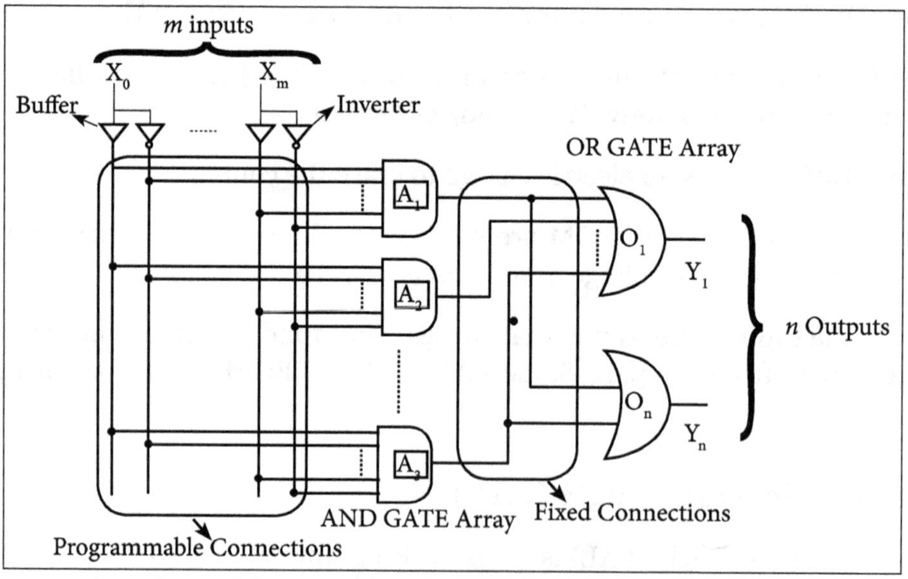

Internal Structure of Programmable Array Logic (PAL).

Further this connection matrix is programmable which lets the user to decide the connection between the input lines and the AND gates. This means that one has to connect each and every input line to either single or multiple AND gate(s), depending on the logic. This causes one to realize the logical 'and' functionality between the input lines.

Further the outputs of the AND gate array are fed as inputs to the OR gates via hard-wired connections, which are fixed and hence unalterable. Moreover it is to be noted that the output of every AND gate is not fed to every OR gate.

For example, OR gate 1 (O_1) has multiple inputs including the outputs of AND gate 1 (A_1), AND gate 2 (A_2) and AND gate p (A_P).

PAL devices consist of a small PROM (Programmable Read-only Memory) core and the additional output logic which is used to implement particular desired logic functions with few components.

Using specialized machines, PAL devices were "field-programmable". PAL's were available in several variants:

"One-time Programmable" (OTP) devices cannot be updated and reused after initial programming (MMI also offered a similar family called HAL or "hard array logic", which were like PAL devices except that they were mask-programmed at the factory).

UV erasable versions (e.g.: PALCxxxxx, e.g.: PALC22V10) have a quartz window over the chip die and could be erased for re-use with an ultraviolet light source just like an EPROM.

Later versions (PALCExxx e.g.: PALCE22V10) were flash erasable devices.

A given column of the OR array has access to only a subset of the possible product terms.

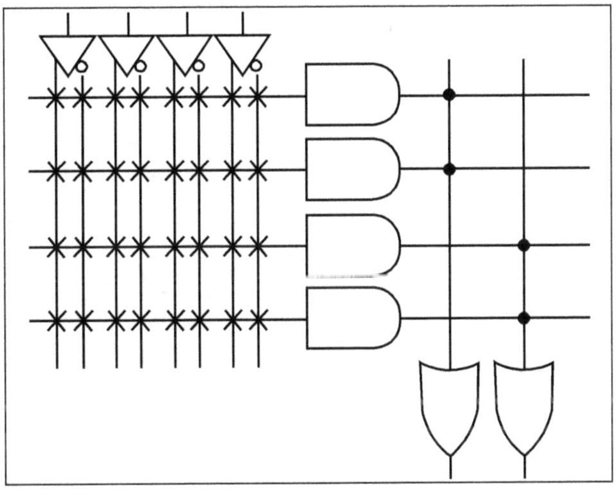

Programmable Logic Array (PLA)

A programmable logic array (PLA) has a programmable AND array at the inputs and programmable OR array at the outputs. The PLA has a programmable AND array instead of hard-wired AND array. The number of AND gates in the programmable AND array are usually much less and the number of inputs of each of the OR gates equals to the number of AND gates.

The OR gate generates an arbitrary Boolean function of minterms equal to the number of AND gates. Figure below shows the PLA architecture with four input lines and a programmable array of eight AND gates at the input and a programmable array of two OR gates at the output.

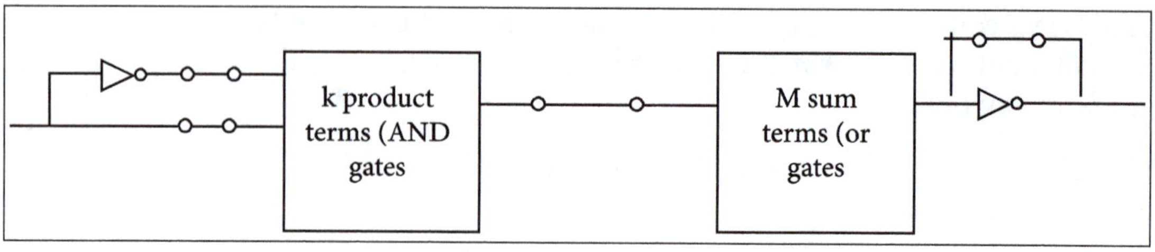

Block Diagram of PLA.

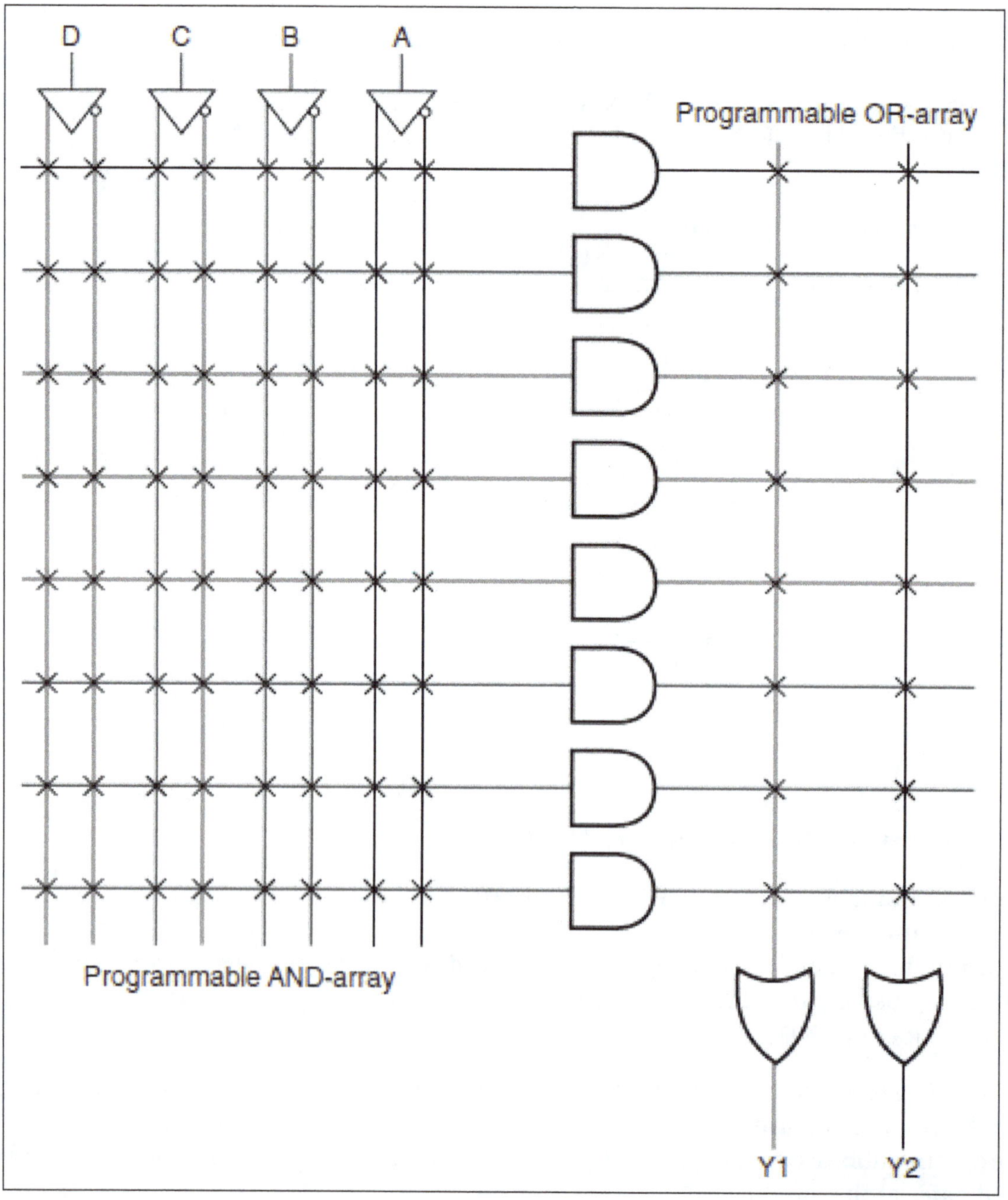

Internal Construction of a Specific PLA

Step 1:

Step 2: The next part is to draw a vertical wire with an AND gate.

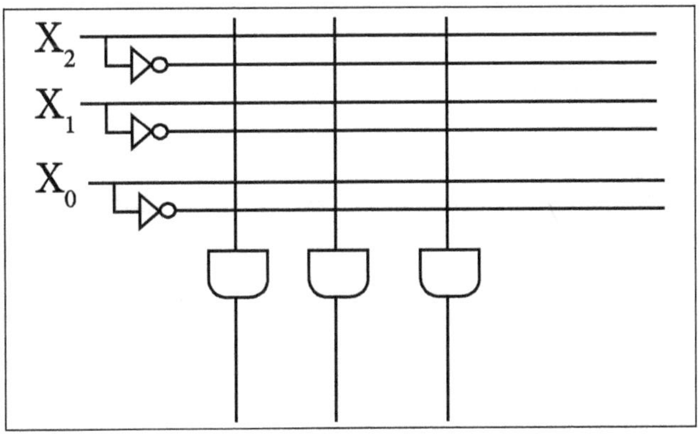

Step 3: The first AND gate (on the left) is the minterm: $X_{2'}.X_{1'}.X_0$.

The second AND gate (from thc lcft) is the minterm: $X_{2'}.X_1.X_0$.

The third AND gate (from the left) is the minterm: $X_{2'}.X_{1'}.X_{0'}$.

We have added a fourth AND gate which is the min term: $X_{2'}.X_{1'}.X_{0'}$.

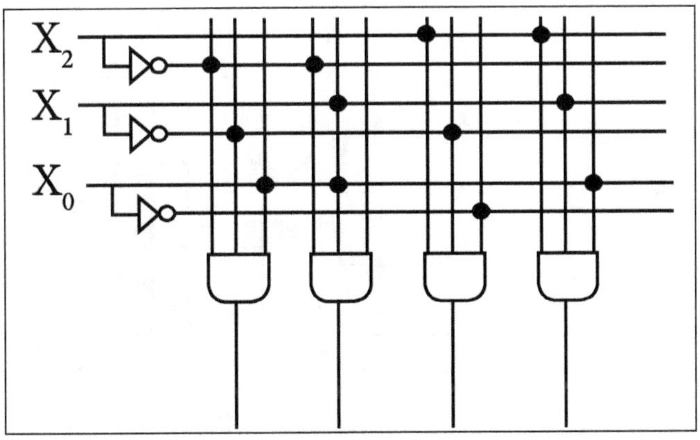

Step 4: We then add OR gates using horizontal wires, to connect the min terms together.

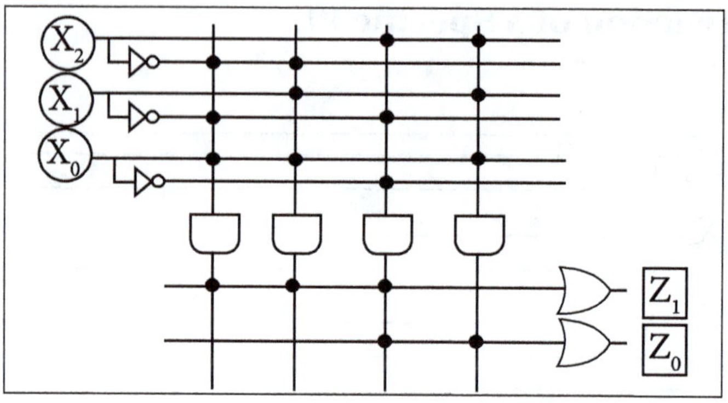

Implementation procedure:

- Preparation in SOP (sum of products) form.
- Obtain the minimum SOP form to reduce the number of product terms to a minimum.
- Decide the input connection of the AND matrix for generating the required product term.
- Then decide the input connections of OR matrix to generate the sum terms.
- Decide the connections of invert matrix.
- Program the PLA.

Let us implement these function f_1 and f_2 as:

$$f_1 = x_1x_2 + x_1\overline{x}_3 + \overline{x}_1\overline{x}_2x_3$$
$$f_2 = x_1x_2 + \overline{x}_1\overline{x}_2\overline{x}_3 + x_1x_3$$

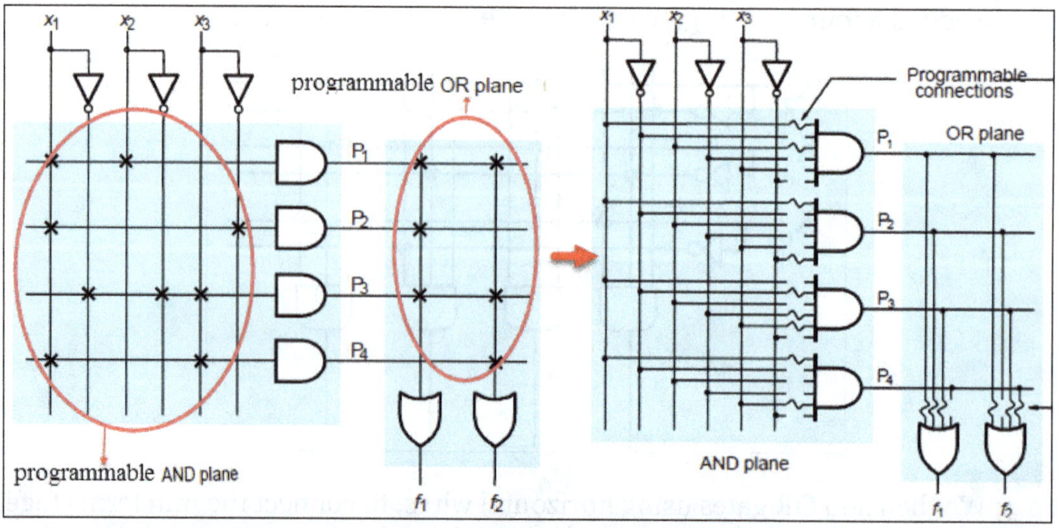

Customary schematic for the PLA Gate-level diagram of a PLA.

Inputs x_1, x_2, x_3 and their respective complemented signals are given to programmable AND plane, therefore we will get AND plane outputs as P_1, P_2, P_3 called minterms. Then these signals are given to programmable OR plane to produce required output function f_1 and f_2 (sum of products).

4.2 Realization of Boolean Function with PLD's

Programmable Logic Devices

A programmable logic device (PLD) is an electronic component used to build reconfigurable digital circuits. Unlike a logic gate, which has a fixed function, a PLD has an undefined function at the time of manufacture. Before the PLD can be used in a circuit it must be programmed, that is, reconfigured.

Programmable Logic Array (PLA)

The PLA is similar to PROM in concept except that PLA does not provide full decoding of the variable and does not generate all the minterms. The decoder is replaced by an array of AND gates that can be programmed to generate any product term of the input variables.

The product terms are then connected to OR gates to provide the sum of products for the required Boolean functions. The internal logic of a PLA with three inputs and two outputs is shown in the below figure:

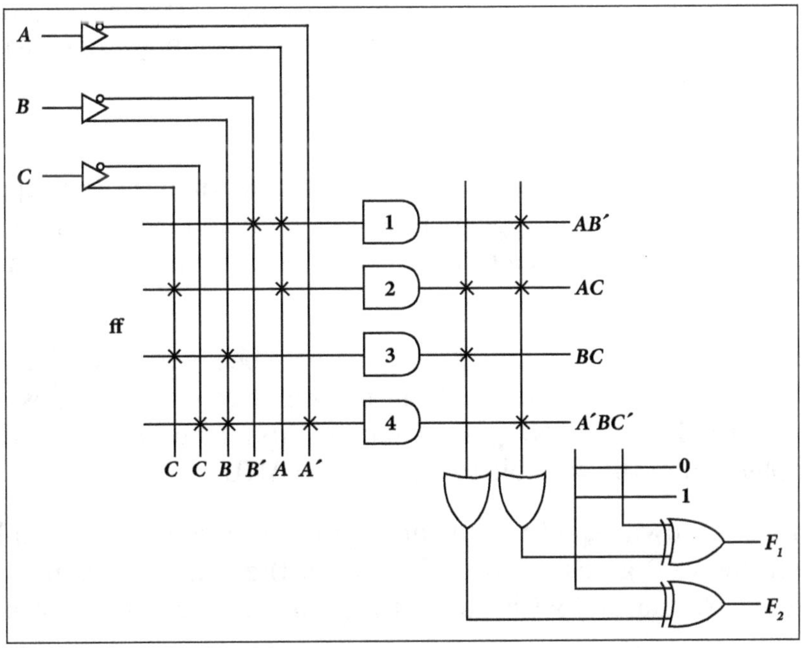

PLA with 3 inputs, 4 product Terms and 2 outputs.

The above diagram uses the array logic graphic symbols for complex circuits. Each input goes through a buffer and an inverter is shown in the above diagram with a composite graphic symbol which has both the true and complement outputs.

Each input and its complement is connected to the inputs of each AND gate as indicated by the intersections between the vertical and horizontal lines. The outputs of the AND gates are connected to the inputs of each OR gate.

The output of the OR gates goes to an XOR gate where the other input can be programmed to receive a signal equal to either logic 1 or 0. The output is inverted when the XOR input is connected to 1. The output does not change when the XOR input is connected to 0. The particular Boolean functions implemented in the PLA is given by:

$$F_1 = AB' + AC + A'BC'$$

$$F_2 = (AC + BC)'$$

The product terms generated in each AND gate is listed along the output of the gate in the above diagram. The product term is determined from the inputs whose cross points are connected and marked with a X. The output of an OR gate gives the logic sum of the selected product terms. The output may be complemented or left in its true form depending on the connection for one of the XOR gate inputs.

The PLA programming table consists of three sections. The first section lists the product terms numerically. The second section specifies the required paths between inputs and AND gates. The third section specifies the paths between AND and OR gates.

For each output variable, we may have a T (for True) or C (for Complement) for programming the XOR gate. The product terms listed on the left are not part of the table; they are included for reference only. For each product term the inputs are marked with 1, 0 or --.

If a variable in the product term appears in its true form, then the corresponding input variable is marked with a 1. If it appears in complemented form, then the corresponding input variable is marked with a 0. If the variable is absent in the product term and then it is marked with a dash.

The paths between the inputs and the AND gates are specified under the column heading inputs in the programming table. A1 in the input column specifies a connection from the input variable to the AND gate. A0 in the input column specifies a connection from the complement of the variable to input of the AND gate.

A dash specifies a blown fuse in both the input variable and its complement. The paths between the AND gates are specified under the column heading outputs. The output variables are marked with 1's for those product terms that are included in the function.

Each product term that has a 1 in the output column requires a path from the output of the AND gate to the input of the OR gate behaves like a 0.

PLA Programming Table

Product Term		Input			Outputs	
				(T)	(C)	
	A	B	C	F1	F2	
AB'	1	1	0	-	1	-
AC	2	1	-	1	1	1
BC	3	-	1	1	-	1
A'BC'	4	0	1	0	1	-

T output dictates that the other input of the corresponding XOR gate be connected to 0, and A C specifies a connection to 1. The size of a PLA is specified by the number of inputs, the number of product terms and the number of outputs.

When designing a digital system with a PLA there is no need to show the internal connections of the units. All that is needed is a PLA programming table from which the PLA can be programmed to supply the required logic.

The PLA may be mask programmable or field programmable with mask programming. The customer submits a PLA program table to the manufacturer. This table is used by the vendor to produce a custom made PLA.

A second type of PLA available is called a field programmable logic array or FPLA. The FPLA can be programmed by the user by means of a commercial hardware programmer unit.

Realization of a Boolean Function using PLA

Implement the following two Boolean functions with a PLA are given by:

$$F_1(A,B,C) = \Sigma\,(0,1,2,4)$$
$$F_2(A,B,C) = \Sigma\,(0,5,6,7)$$

The two functions are simplified as shown in maps. Both the true and complement of the functions are simplified in sum of products.

The combination that gives a minimum number of product terms is:

$$F_1 = (AB+AC+BC)'$$
$$F_2 = AB+AC+A'B'C'$$

This gives four district product terms AB, AC, BC and A'B'C'.

The PLA programming table for this combination is shown in the below table:

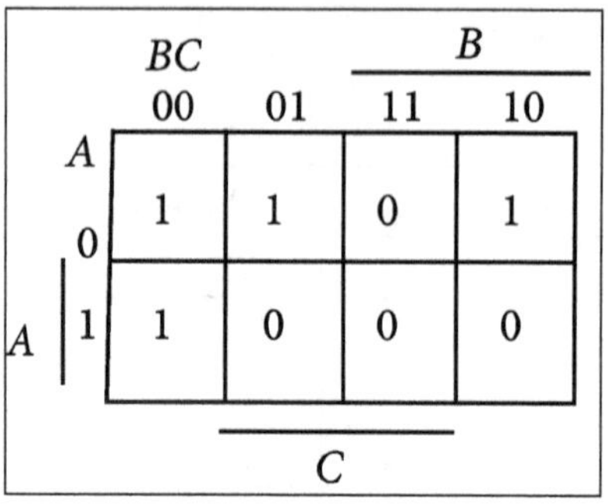

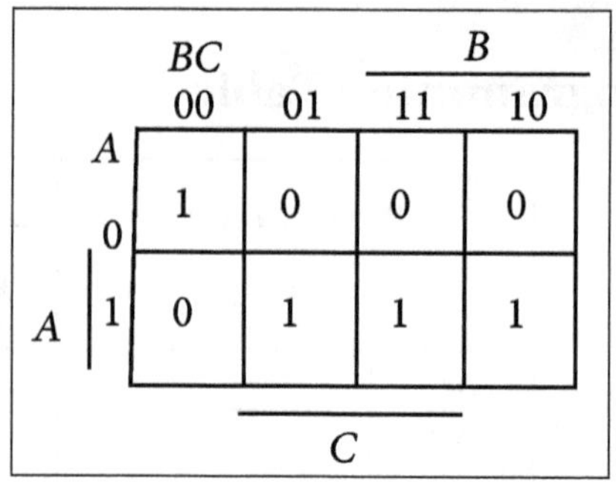

$$F_1 = A'B' + A'C' + B'C'$$

$$F_1 = (AB + AC + BC)'$$

$$F_2 = AB + AC + A'B'C'$$

$$F_2 = (A'C + A'B + AB'C')'$$

PLA programming table						
	Product Term	Input			Outputs	
		A	B	C	(C) F_1	(T) F_2
AB	1	1	1	-	1	1
AC	2	1	-	1	1	1
BC	3	-	1	1	1	-
A'BC'	4	0	0	0	-	1

It is important that output F_1 is the true output, even though a c is marked over it in the table. This is because F_1 is generated with an AND-OR circuit and is available at the output of the OR gate. The XOR gate complements the function to produce the true F_1 output.

The combinational circuit used in this example is too simple for implementing with a PLA.

Programmable Array Logic (PAL)

The programmable array logic (PAL) is a programmable logic device with a fixed OR array and a programmable AND array. Since the AND gates are programmable, the PAL is easier to program but is not as flexible as the PLA.

It has four inputs and four outputs. Each input has a buffer inverter gate and each output is generated by a fixed OR gate. There are four sections in the unit, each being composed of a three wide AND-OR array.

Each AND gate has 10 programmable input connections. This is shown in the diagram by 10 vertical lines intersecting each horizontal line. The horizontal line symbolizes the multiple input configuration of the AND gate. One of the outputs is connected to a buffer inverter gate and then fed back into two inputs of the AND gates.

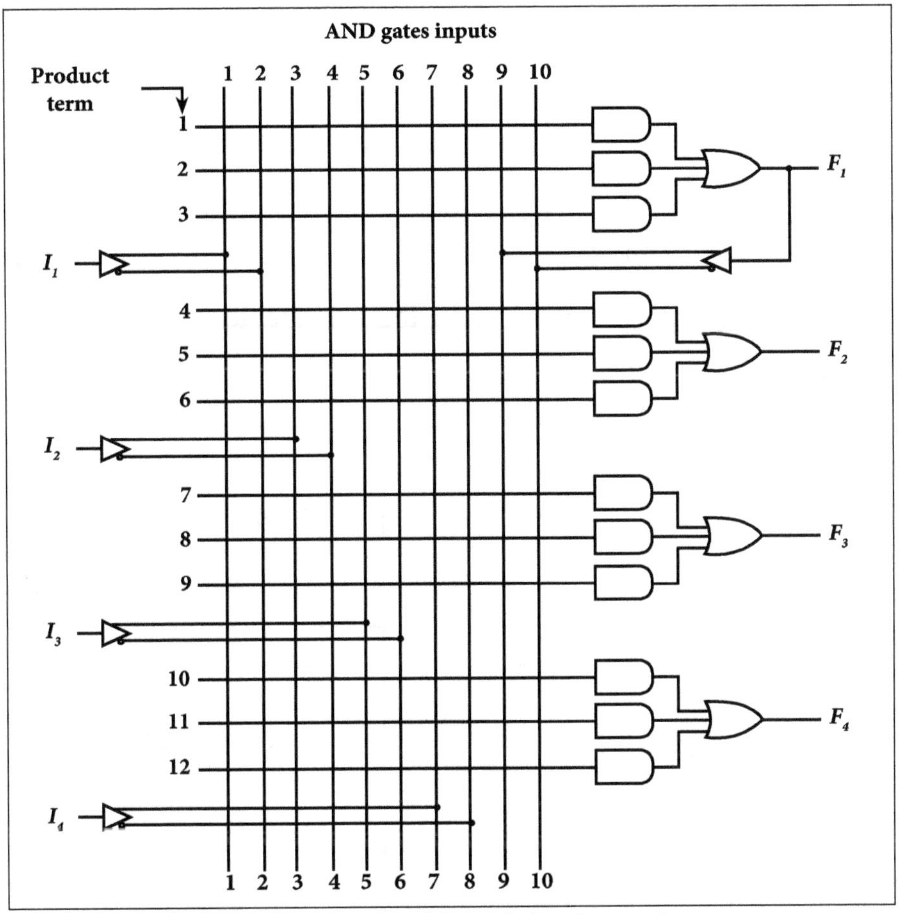

Commercial devices contain PAL with four inputs, four outputs and three wide AND–OR structure.

When designing with a PAL, the Boolean functions must be simplified to fit into each section. Unlike PLA, a product term cannot be shared among two or more OR gates. Therefore each function can be simplified by itself without regard to common product terms.

Realization of a Combinational Circuit using PAL

Consider the following Boolean functions given in sum of minterms:

$$W(A,B,C,D) = \Sigma(2,12,13)$$
$$X(A,B,C,D) = \Sigma(7,8,9,10,11,12,13,14,15)$$
$$Y(A,B,C,D) = \Sigma(0,2,3,4,5,6,7,8,10,11,15)$$
$$Z(A,B,C,D) = \Sigma(1,2,8,12,13$$

Simplify the four functions to a minimum number of terms which result in the following Boolean functions:

$$W = ABC' + A'B'CD'$$

$$X = A + BCD$$

$$Y = A'B + CD + B'D'$$

$$Z = ABC' + A'B'CD' + AC'D' + A'B'C'D = W + AC'D' + A'B'C'D.$$

The function for Z has four product terms. The logical sum of two of these terms is equal to W. By using W, it is possible to reduce the number of terms for Z from four to three.

The PAL Programming Table is shown in the following table:

Product term	AND Inputs					Outputs
	A	B	C	D	W	
1 2 3	1	1	0	-	-	W = ABC' + A'B'CD'
	0	0	1	0	-	
	-	-	-	-	-	
4 5 6	1	-	-	-	-	X = A + BCD
	-	1	1	1	-	
	-	-	-	-	-	
7 8 9	0	1	-	-	-	y = A'B + CD + B'D'
	-	-	1	1	-	
	-	0	-	0	-	
10 11 12	-	-	-	-	1	Z = w + AC'D' + A'B'C'D
	1	-	0	0	-	
	0	0	0	1	-	

The table is divided into four sections with three product terms. The first two sections need only two product terms to implement the Boolean function. The last section for output Z needs four product terms. Using the output from W, we can reduce the function to three terms.

The fuse map for the PAL as specified in the programming table is shown in the below figure:

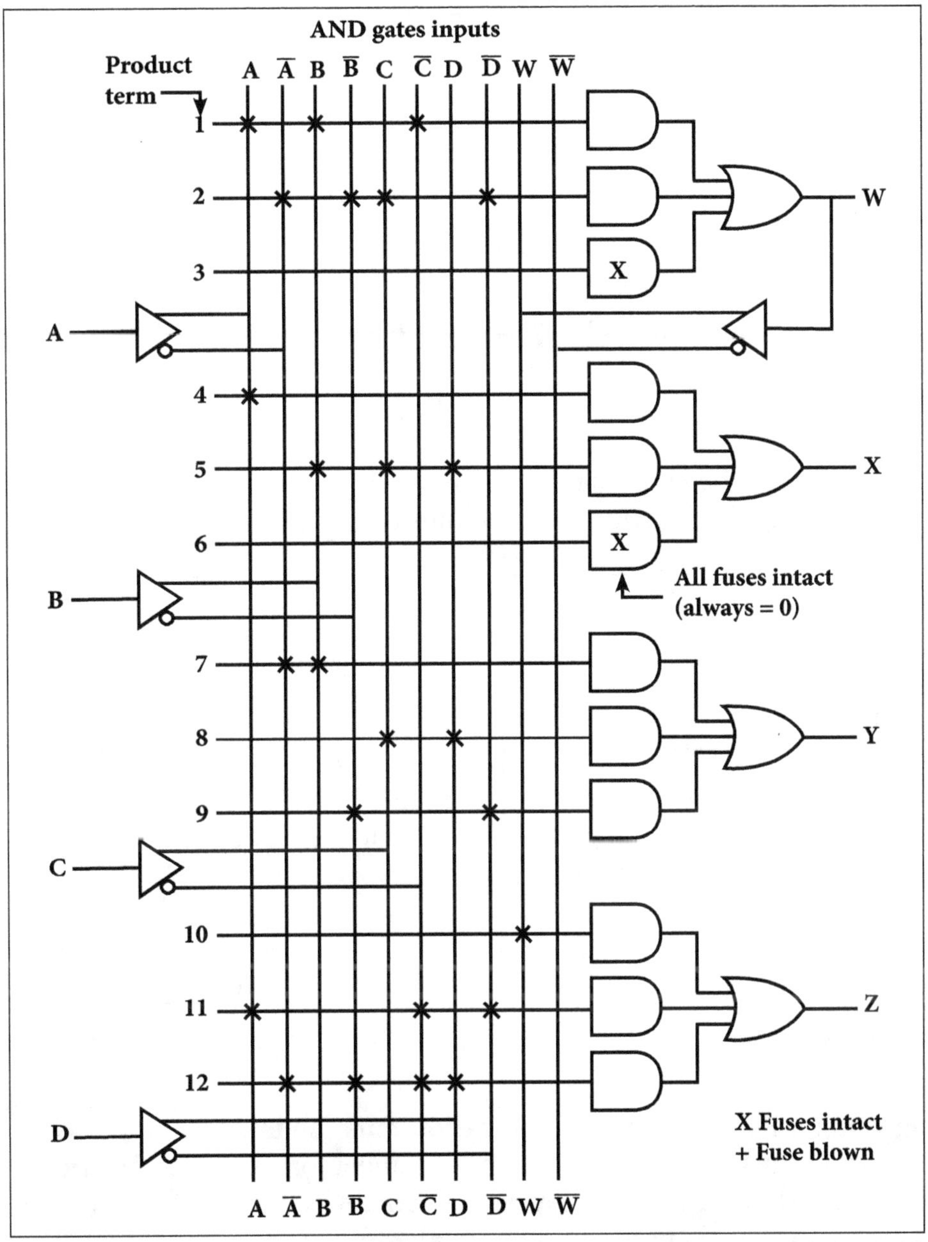

For each 1 or 0 in the table, we mark the corresponding intersection in the diagram with the symbol for an intact fuse. For each dash we mark the diagram with blown fuses in both the true and complement inputs. If the AND gate is not used, we leave all its input fuses intact.

Since, the corresponding input receives both the true and complement of each variable, we have AA' = 0 and the output of the AND gate is always 0.

Example: Implement the combinational circuit having the shown truth table, using PLA.

A	B	C	F_1	F_2
0	0	0	1	1
0	0	1	1	0
0	1	0	1	0
0	1	1	0	0
1	0	0	1	0
1	0	1	0	1
1	1	0	0	1
1	1	1	0	1

Each product term in the expression requires an AND gate. To minimize the cost, it is necessary to simplify the function to a minimum number of product terms.

A \ BC	00	01	11	10
0	1	1	0	1
1	1	0	0	0

A \ BC	00	01	11	10
0	1	0	0	0
1	0	1	1	1

$$F_1 = \overline{A}\,\overline{B} + \overline{A}\,\overline{C} + \overline{B}\,\overline{C}$$
$$\overline{F}_1 = AB + AC + BC$$

$$F_2 = AB + AC + \overline{A}\,\overline{B}\,\overline{C}$$
$$\overline{F}_2 = \overline{A}C + \overline{A}B + \overline{A}\,\overline{B}\,\overline{C}$$

Designing using a PLA, a careful investigation must be taken in order to reduce the distinct product terms. Both the true and complement forms of each function should be simplified to see which one can be expressed with fewer product terms and which one provides product terms that are common to other functions.

The combination that gives a minimum number of product terms is:

$F_1' = AB + AC + BC$ or $F_1 = (AB + AC + BC)'$
$F_2 = AB + AC + A'B'C'$

This gives only 4 distinct product terms: AB, AC, BC and A'B'C'.

So the PLA table will be as follows:

PLA programming table						
	Product Term	Input			Outputs	
		A	B	C	(C) F1	(T) F2
AB	1	1	1	-	1	1
AC	2	1	-	1	1	1
BC	3	-	1	1	1	-
$\overline{A}\overline{B}\overline{C}$	4	0	0	0	-	1

For each product term, the inputs are marked with 1, 0, or – (dash). If a variable in the product term appears in its normal form (unprimed), the corresponding input variable is marked with 1. A1 in the input column specifies a path from the corresponding input to the input of the AND gate that forms the product term.

A0 in the input column specifies a path from the corresponding complemented input to the input of the AND gate. A dash specifies no connection.

The appropriate fuses are blown and the ones left intact form the desired paths. It is assumed that the open terminals in the AND gate behave like a 1 input. In the output column, T (True) specifies that the other input of the corresponding XOR gate can be connected to 0 and C (Complement) specifies a connection to 1.

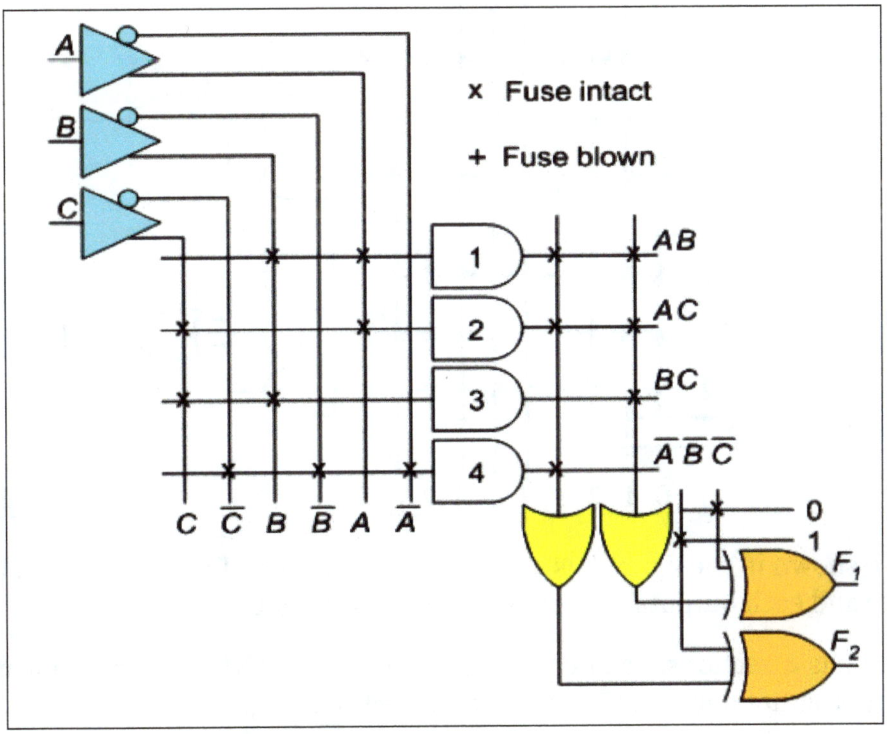

It is important that output F_1 is the normal (or true) output even though C (for complement) is marked over it. This is because F_1' is generated with AND-OR circuit prior to the output XOR. The output XOR complements the function F_1' to produce the true F_1 output as its second input is connected to logic 1.

The PAL (Programmable Array Logic)

The PAL device is a PLD with a fixed OR array and a programmable AND array.

As only AND gates are programmable, the PAL device is easier to program but it is not as flexible as the PLA.

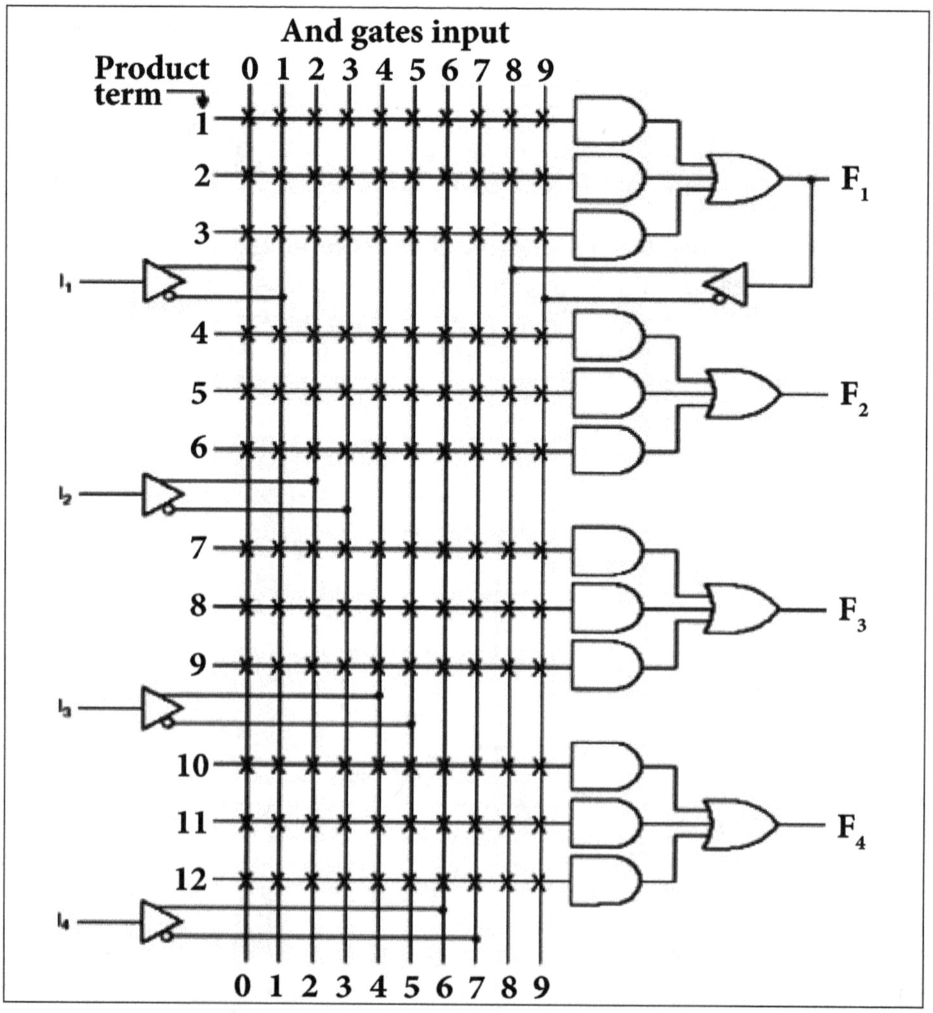

The device shown in the figure has 4 inputs and 4 outputs. Each input has a buffer inverter gate and each output is generated by a fixed OR gate.

The device has 4 sections, each composed of a 3-wide AND-OR array, meaning that there are 3 programmable AND gates in each section.

Each AND gate has 10 programmable input connections indicating by 10 vertical lines intersecting each horizontal line. The horizontal line symbolizes the multiple input configuration of an AND gate.

One of the outputs F_1 is connected to a buffer-inverter gate and is fed back into the inputs of the AND gates through programmed connections.

Designing using a PAL device, the Boolean functions must be simplified to fit into each section. The number of product terms in each section is fixed and if the number of terms in the function is too large, it may be necessary to use two or more sections to implement one Boolean function.

Example

Implement the following Boolean functions using the PAL device as shown above:

$$W(A, B, C, D) = \Sigma m(2, 12, 13)$$
$$X(A, B, C, D) = \Sigma m(7, 8, 9, 10, 11, 12, 13, 14, 15)$$
$$Y(A, B, C, D) = \Sigma m(0, 2, 3, 4, 5, 6, 7, 8, 10, 11, 15)$$
$$Z(A, B, C, D) = \Sigma m(1, 2, 8, 12, 13)$$

Simplifying the 4 functions to a minimum number of terms results in the following Boolean functions:

W = ABC' + A'B'CD'

X = A + BCD

Y = A'B + CD + B'D'

Z = ABC' + A'B'CD + AC'D' + A'B'C'D = W + AC'D' + A'B'C'D

It is important that the function for Z has four product terms. The logical sum of two of these terms is equal to W. Thus, by using W, it is possible to reduce the number of terms for Z from four to three, so that the function can fit into the given PAL device.

The PAL programming table is similar to the table used for the PLA, except that only the inputs of the AND gates need to be programmed.

Product Term	AND Inputs					Outputs
	A	B	C	D	W	
1	1	1	0	–	–	$W = AB\overline{C}$
2	0	0	1	0	–	$+ \overline{A}\overline{B}C\overline{D}$
3	–	–	–	–	–	
4	1	–	–	–	–	$X = A$
5	–	1	1	1	–	$+ BCD$
6	–	–	–	–	–	

7	0	1	–	–	–	$Y = \bar{A}B$
8	–	–	1	1	–	$+CD$
9	–	0	–	0	–	$+\bar{B}\bar{D}$
10	–	–	–	–	1	$Z = W$
11	1	–	0	0	–	$+A\bar{C}\bar{D}$
12	0	0	0	1	–	$+\bar{A}\bar{B}\bar{C}D$

The figure shows the connection map for the PAL device, as specified in the programming table:

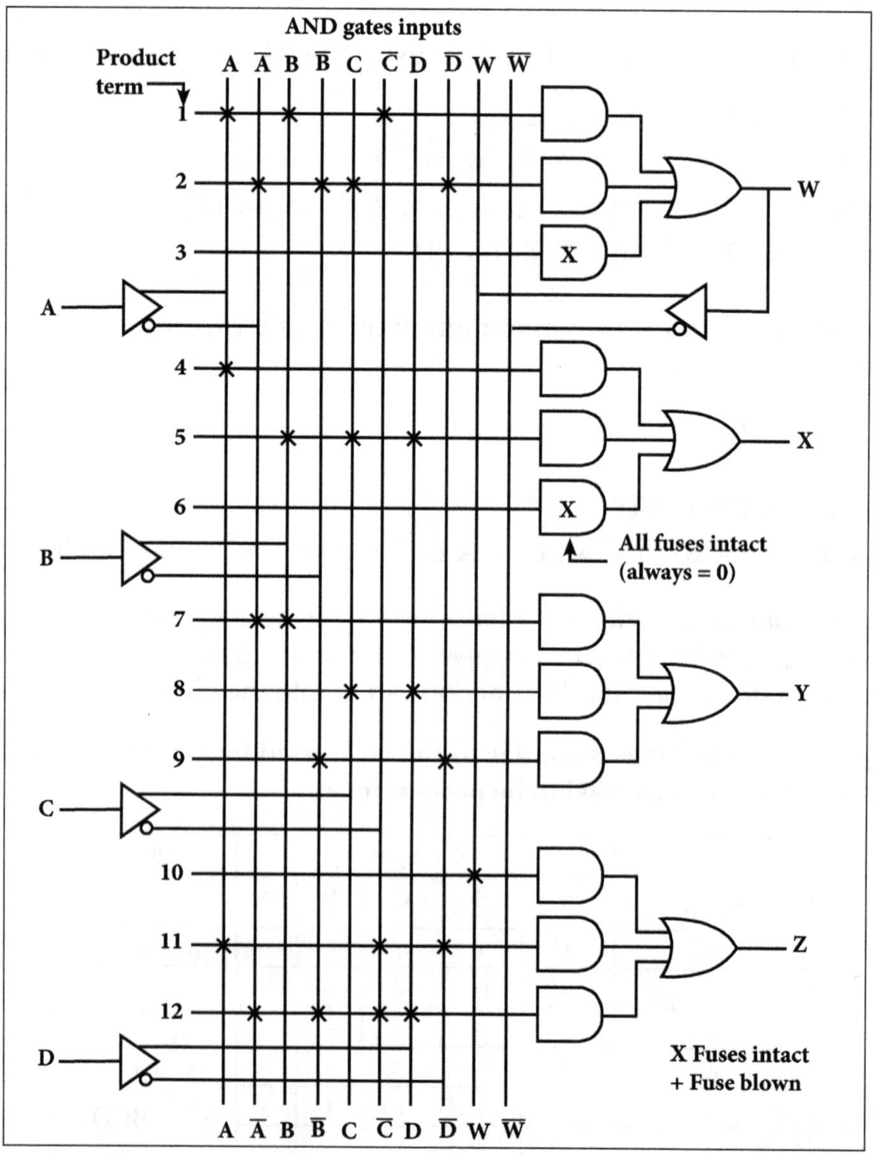

Since both W and X have two product terms, third AND gate is not used. If all the inputs to this AND gate are left intact, then its output will always be 0, because it receives both the true and complement of each input variable i.e., AA' = 0.

4.3 Programming Tables of PLD's

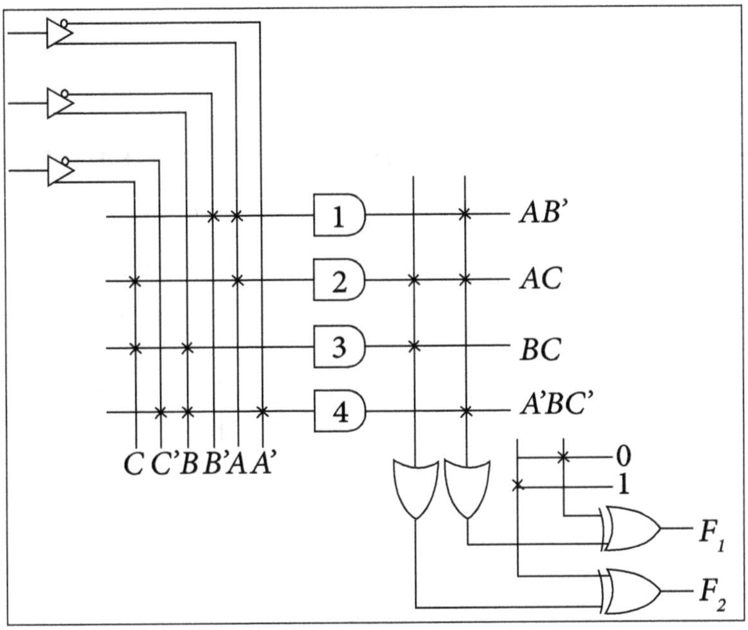

Programmable logic device.

PLA Programming Table

Product	Inputs			Outputs	
term	A	B	C	F_1	F_2
1	1	0	-	1	-
2	1	-	1	1	1
3	-	1	1	-	1
4	0	1	0	1	-

Programming Table

- First: lists the product terms numerically.
- Second: specifies the required paths between inputs and AND gates.
- Third: specifies the paths between the AND and OR gates.

- For each output variable, we may have a T (True) or C (Complement) for programming the XOR gate.

PAL Table

- Z has four product terms and we can replace by W with two product terms, this will reduce the number of terms for Z from four to three.

Product Term	AND Inputs					Outputs
	A	B	C	D	W	
1	1	1	0	–	–	w=ABC'
2	0	0	1	0	–	+A'B'C'D'
3	–	–	–	–	–	
4	1	–	–	–	–	x=A
5	–	1	1	1	–	+BCD
6	–	–	–	–	–	
7	0	1	–	–	–	y=A'B
8	–	–	1	1	–	+CD
9	–	0	–	0	–	+B'D'
10	–	–	–	–	1	z=w
11	1	–	0	0	–	+AC'D'
12	0	0	0	1	–	+A'B'C'D

4.4 Merits and Demerits of PROM, PAL and PLA Comparison

S. No.	PROM	PLA	PAL
1.	In PROM, AND array is fixed and OR array is programmable.	Both AND & OR arrays are programmable.	AND array is programmable and OR array is fixed.
2.	Cheaper and Simpler	Costliest & complex than PAL & ROMS.	Cheaper and simpler.
3.	All minterms are decoded.	AND array and OR array can be programmed to get desired minterms.	AND array can be programmed to get desired minterms.
4.	Only Boolean functions in canonical SOP form can be implemented using PROM.	Any Boolean function in standard SOP form can be implemented using PLA.	Any Boolean function in standard SOP form can be implemented using PAL.

4.5 Realization of Boolean Functions using PROM

A PROM is characterized by the number of output lines of the decoder and the number of output lines from the or-array. The structure of PROM is given in the below figure. The realization of a set of Boolean expressions using a decoder and or-gates is equivalent to the very same approach using a PROM. A PROM is a device that includes both the decoder and or-gates within the same network.

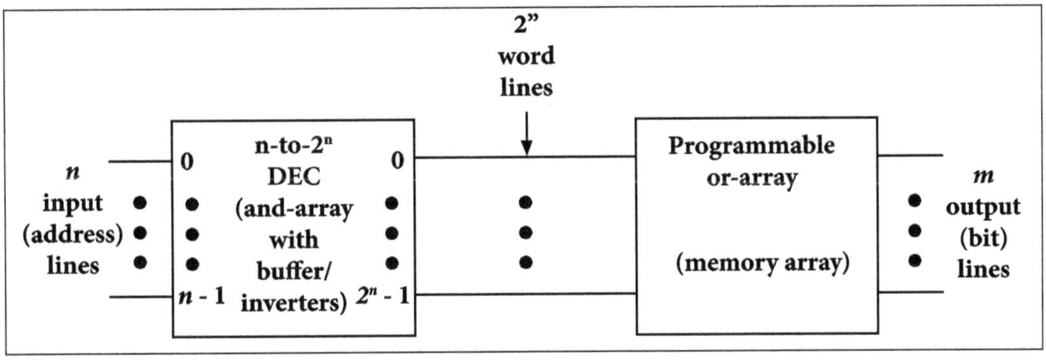

Structure of PROM.

Given a set of Boolean expressions in minterm canonical form or a set of Boolean functions in truth table form, it is only necessary to determine which programmable links of a PROM to retain and which to open.

The programming of the PROM is then carried out by blowing the appropriate fuses. PROMS are typically used for code conversions, generating bit patterns for characters and as lookup tables for arithmetic functions.

As a simple example of using a PROM for combinational logic design. Consider the Boolean expressions as follows:

$$f_1(x_2, x_1, x_0) = \sum m(0,1,2,5,7)$$

$$f_2(x_2, x_1, x_0) = \sum m(1,2,4,6)$$

Since, these are functions of three input variables, a PROM having a 3-to-8-line decoder is needed. In addition, since there are two functions being realized, the or-array must consist of two gates. Hence, an 8x2 PROM is needed for the realization.

A blown fusible link on the input of an or-gate is equivalent to a logic-0 input. It should be emphasized that this example is for illustrative purposes only. From a practical point of view, PROM's are intended for combinational networks having a large number of inputs and outputs.

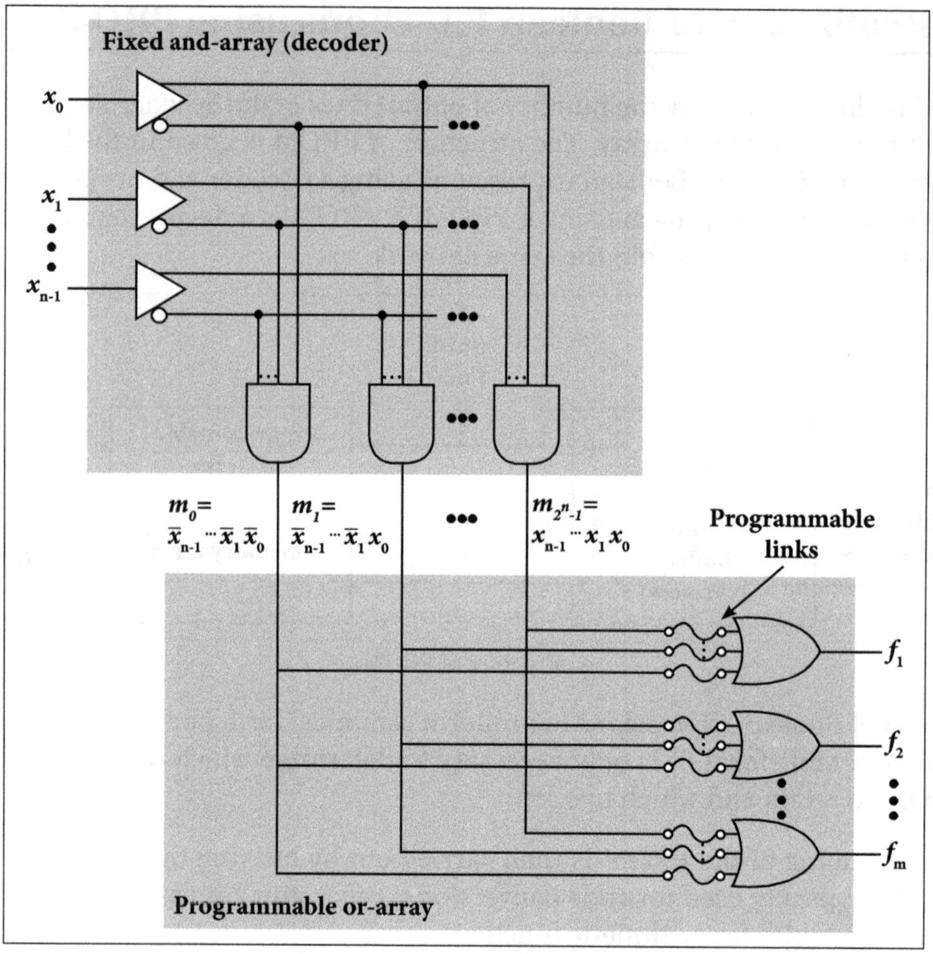

(a) A 2n*m PROM Logic Diagram.

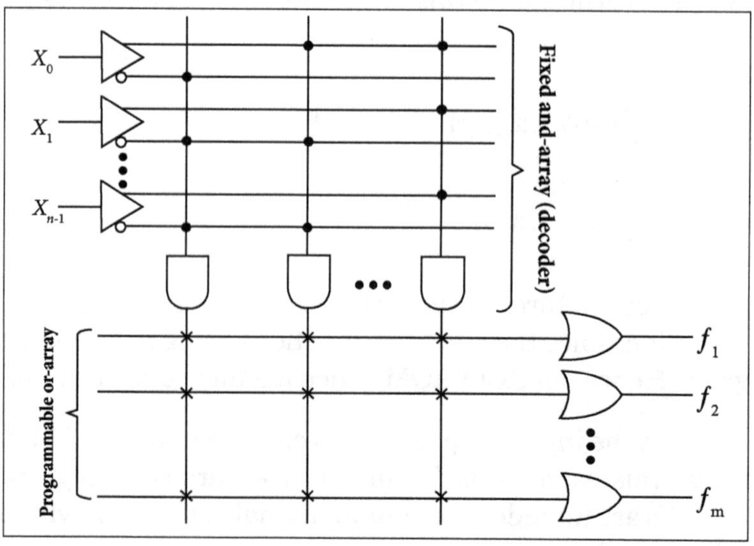

(b) Representation in PLD notation.

x_2	x_1	x_0	f_1	f_2
0	0	0	1	0
0	0	1	1	1
0	1	0	1	1
0	1	1	0	0
1	0	0	0	1
1	0	1	1	0
1	1	0	0	1
1	1	1	1	0

Truth table.

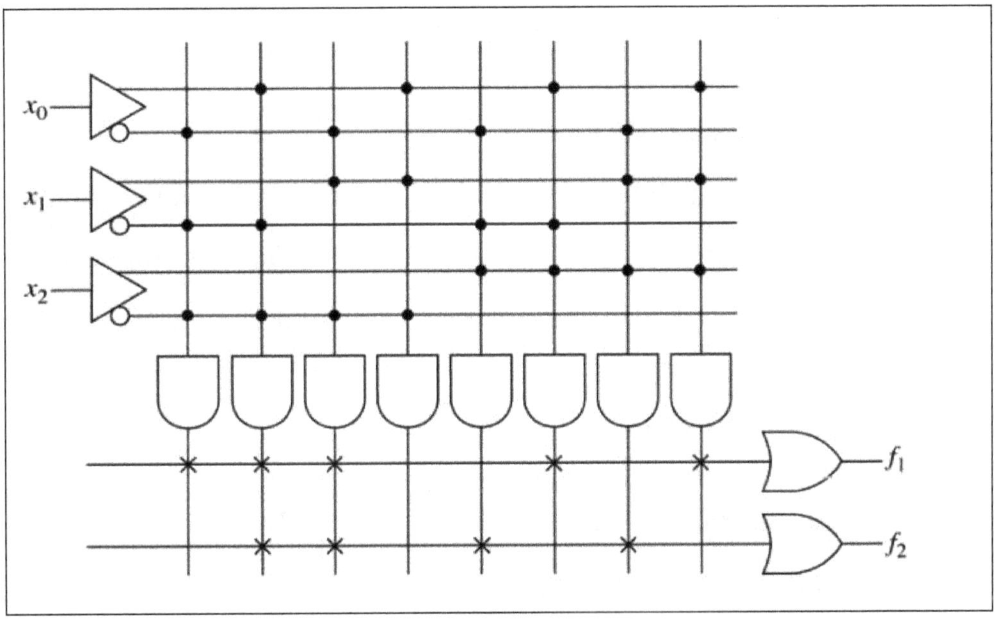

PROM realization.

It may seem strange that the structure of PROM as a logic-design device is called a read-only memory. Read-only memory devices were originally developed to store permanent data in a digital system. In these devices, each piece of data called a word is accessible by specifying an address.

By applying a 3-bit combination to the x_0, x_1 and x_2 lines, precisely one 'and' gate in the decoder is selected in the sense that its output line. i.e., word line, is logic-1. Thus, each input combination is regarded as an address of one of the word lines.

As a consequence of selecting a given word line, a pattern of 0's and 1's i.e., a word, as determined by the fusible connections to the selected word line appears at the output terminals. For example, the bit lines, of the device. This 0-1 pattern is considered the word stored at the address associated with the selected word line.

For example, the word stored at address $x_2x_1x_0$ = 100 in the above figure is f_1f_2 = 01. Finally, the fact that the connections associated with the fusible links normally cannot be altered once they are formed makes the term read-only appropriate for this device. Hence, the realization shown in the above figure is a read-only memory storing four words each consisting of 2 bits.

For each additional input line to a PROM, the number of gates in the decoder and the number of inputs to each gate in the or-array double. This is because all possible min terms are generated by the decoder and all the min terms appear as inputs to the gates in the or-array. However, in many applications, not all the min terms are necessary. In such cases, the and-array is not utilized efficiently.

Also, we know that the collections of min terms can frequently be re-placed by a single product term. If the and-array is made programmable so that only necessary product terms are generated, then its size can be controlled.

Programmable Array Logic (PAL)

- A compact form of the internal logic of PLD's can be referred to as array logic.
- When designing with a PAL, the Boolean functions must be simplified.
- Unlike the PLA, a product term cannot be shared among two or more OR gates.

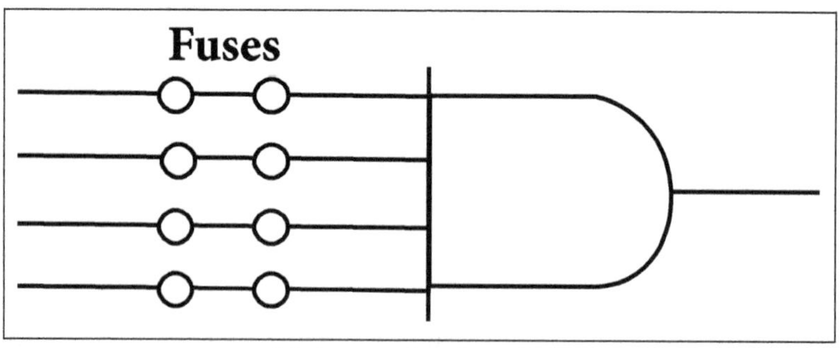

(a) Conventional symbol.

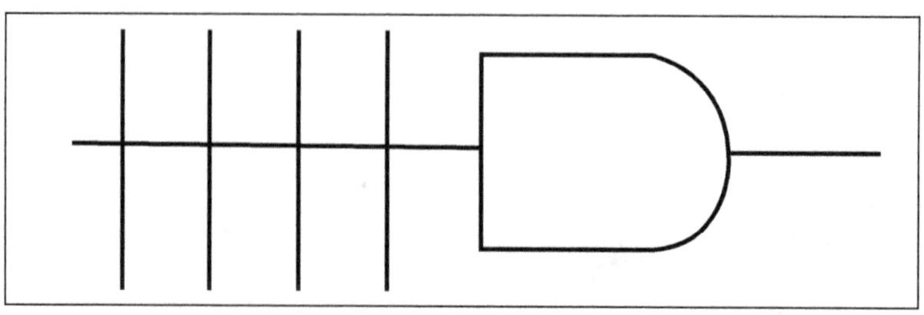

(b) Array logic symbol.

Example: Considering the following Boolean functions, design a combinational circuit using a PAL circuit design.

$$w(A,B,C,D)=\sum(2,12,13)$$

$$x(A,B,C,D)=\sum(7,8,9,10,11,12,13,14,15)$$

$$y(A,B,C,D)=\sum(0,2,3,4,5,6,7,8,10,11,15)$$

$$z(A,B,C,D)=\sum(1,2,8,12,13)$$

Solution:

- Simply the four functions to a minimum number of terms:

$$w = ABC' + A'B'CD'$$
$$x = A + BCD$$
$$y = A'B + CD + B'D'$$
$$z = ABC' + A'B'C'D + AC'D' + A'B'C'D$$
$$= w + AC'D' + AB'C'D$$

- Its Truth table is given in the following table:

Product Term	AND Inputs					Outputs
	A	B	C	D	W	
1	1	1	0	–	–	w=ABC' +A'B'CD'
2	0	0	1	0	–	
3	–	–	–	–	–	
4	1	–	–	–	–	x=A +BCD
5	–	1	1	1	–	
6	–	–	–	–	–	
7	0	1	–	–	–	y=A'B +CD +B'D'
8	–	–	1	1	–	
9	–	0	–	0	–	
10	–	–	–	–	1	z=w +AC'D' +A'B'C'D
11	1	–	0	0	–	
12	0	0	0	1	–	

- Logic Diagram is given in the figure:

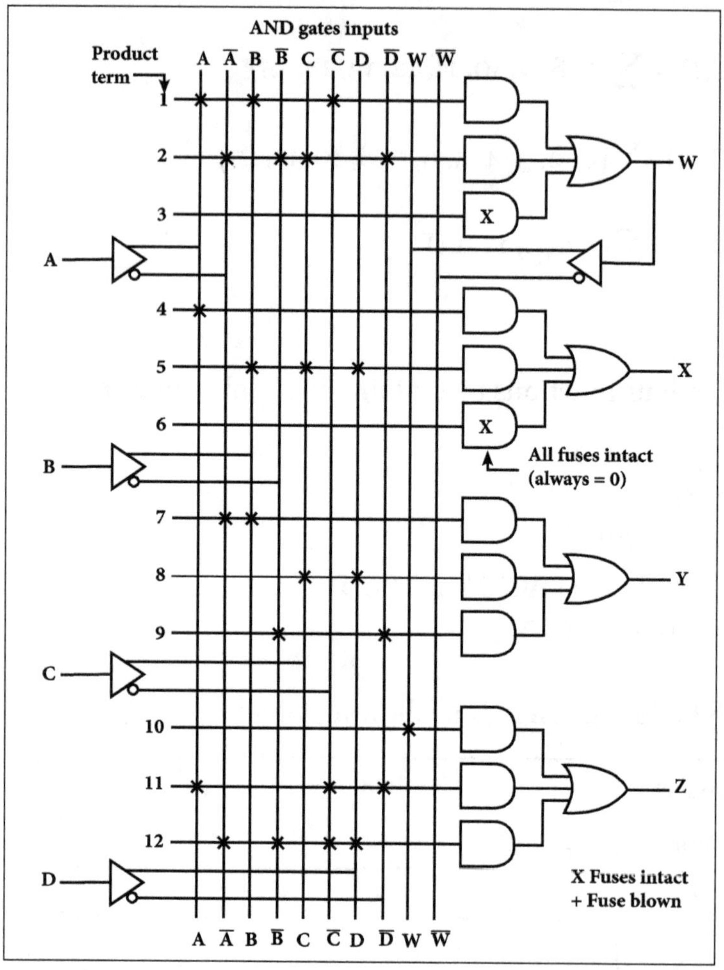

Problems

1. Let us design a large circuit for the following Boolean expressions using PROM:

$$F_{1(x,y,z)} = \sum m(1, 2, 4, 7)$$

$$F_{2(x,y,z)} = \sum m(3, 5, 6, 7)$$

Solution:

Given:

$$F_{1(x,y,z)} = \sum m(1, 2, 4, 7)$$

$$F_{2(x,y,z)} = \sum m(3, 5, 6, 7)$$

The Truth Table for the functions F_1 and F_2 are given as:

Inputs			Outputs	
X	Y	Z	F1	F2
0	0	0	0	0
0	0	1	1	0
0	1	0	1	0
0	1	1	0	1
1	0	0	1	0
1	0	1		1
1	1	0	0	1
1	1	1	1	1

The PROM implementation is given below as:

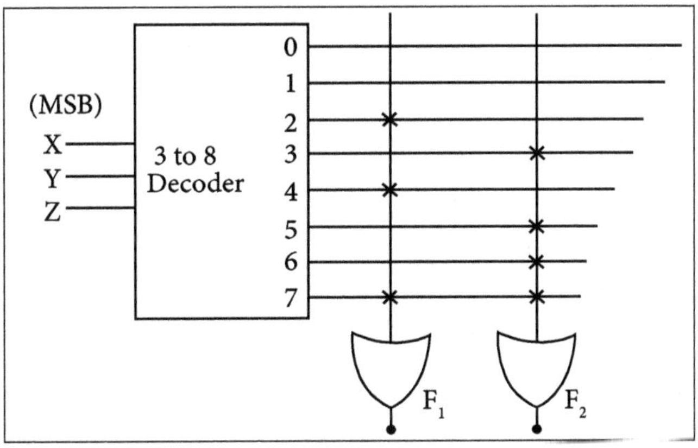

4 - bit binary to Gray code converter using PLA.

Binary				Gray			
B3	B2	B1	B0	G3	G2	G1	G0
0	0	0	0	0	0	0	0
0	0	0	1	0	0	0	1
0	0	1	0	0	0	1	1
0	0	1	1	0	0	1	0
0	1	0	0	0	1	1	0
0	1	0	1	0	1	1	I
0	1	1	0	0	1	0	1
0	1	1	1	0	1	0	0
1	0	0	0	1	1	0	0
1	0	0	1	1	1	0	1
1	0	1	0	1	1	1	1
1	0	1	1	1	1	1	0
1	1	0	0	1	0	1	0

1	1	0	1	1	0	1	1
1	I	1	0	1	0	0	1
1	1	1	1	1	0	0	0

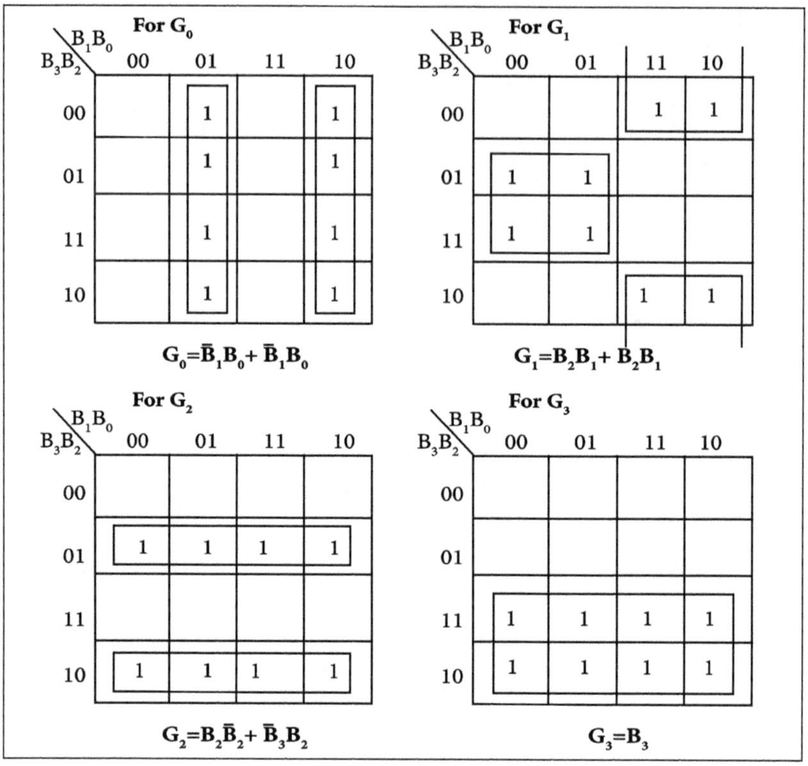

K-Map.

Product	Inputs				Outputs			
Term	B_3	B_2	B_1	B_0	G_3	G_2	G_1	G_0
$B_1\bar{B}_0$	-	-	1	0	-	-	-	1
$\bar{B}_1B_0$	-	-	0	1	-	-	-	1
$\bar{B}_2B_1$	-	0	1	-	-	-	1	-
$B_2\bar{B}_1$	-	1	0	-	-	-	1	-
$\bar{B}_3B_2$	1	0	-	-	-	1	-	-
$B_3\bar{B}_2$	1	0	-	-	-	1	-	-
B_3	1	-	-	-	1	-	-	-
					T	T	T	T/C

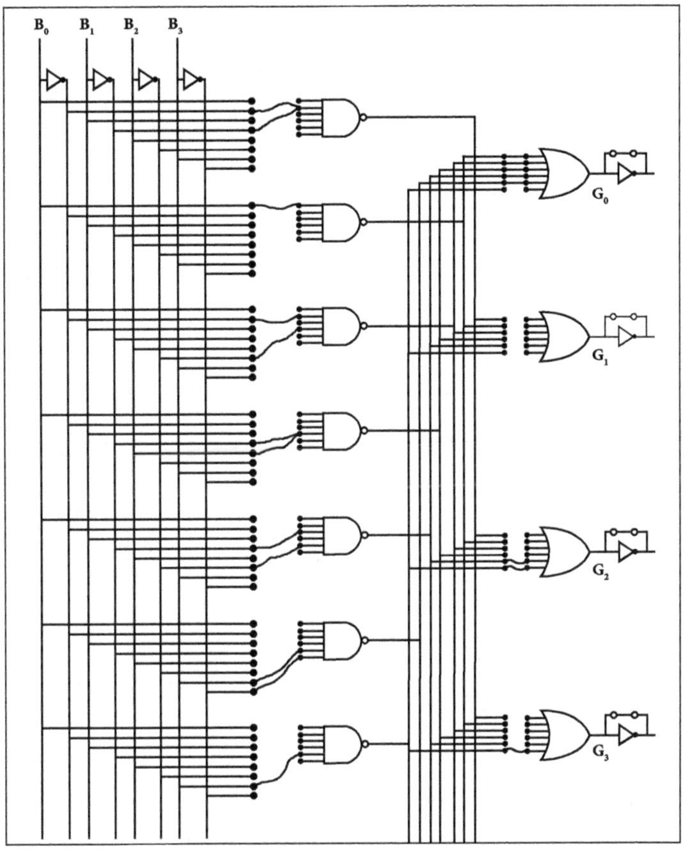

Logic diagram.

2. Let us implement the following function using PLA:

A	B	C	F_1	F_1
0	0	0	0	0
0	0	1	0	0
0	1	0	0	0
0	1	1	0	1
1	0	0	1	0
1	0	1	1	1
1	1	0	0	0
1	1	1	1	1

Solution:

Given:

A	B	C	F_1	F_1
0	0	0	0	0
0	0	1	0	0

0	1	0	0	0
0	1	1	0	1
1	0	0	1	0
1	0	1	1	1
1	1	0	0	0
1	1	1	1	1

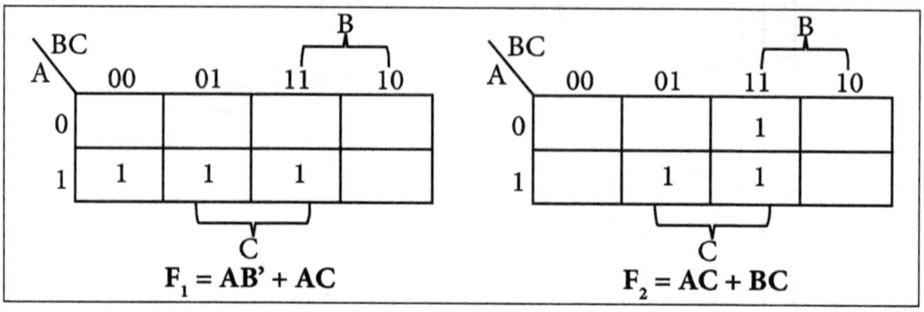

K–Map.

Truth Table

	Product Term	Inputs			Outputs		
AB'	1	1	0	-	1	-	
AC'	2	1	-	1	1	1	
BC'	3	-	1	1	-	1	
					T	T	T/C

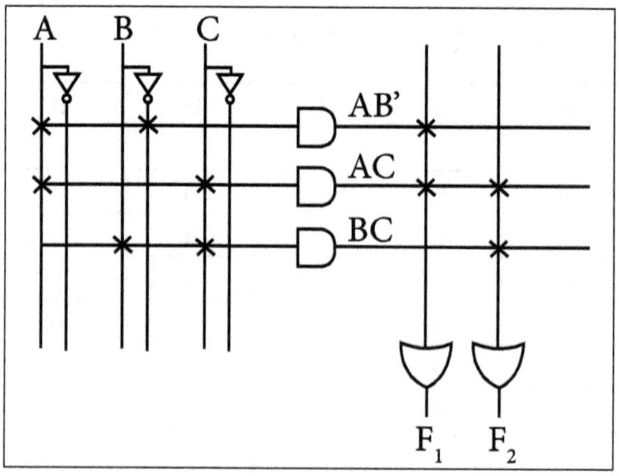

Logic diagram.

3. Let us implement the following two Boolean functions with a PLA:

$$F_1(A,B,C) = \sum(0,1,2,4)$$

$$F_2(A,B,C) = \sum(0,5,6,7)$$

Solution:

Given:

$$F_1(A,B,C)=\sum(0,1,2,4)$$

$$F_2(A,B,C)=\sum(0,5,6,7)$$

$$F_1(A,B,C)=\bar{A}\bar{B}\bar{C}+\bar{A}\bar{B}C+\bar{A}B\bar{C}+A\bar{B}\bar{C}$$

$$F_2(A,B,C)=\bar{A}\bar{B}\bar{C}+AC+AB$$

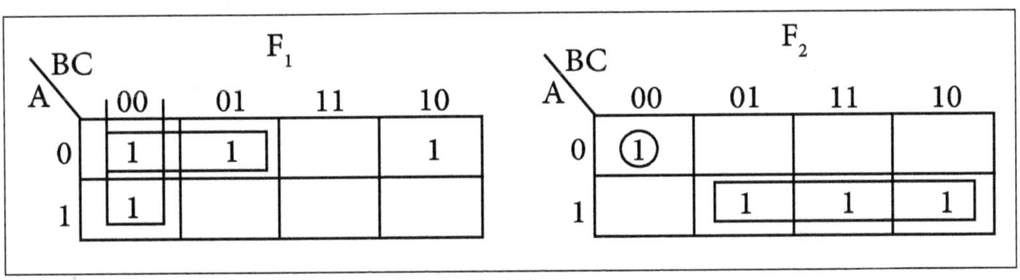

K–Map.

Implementation

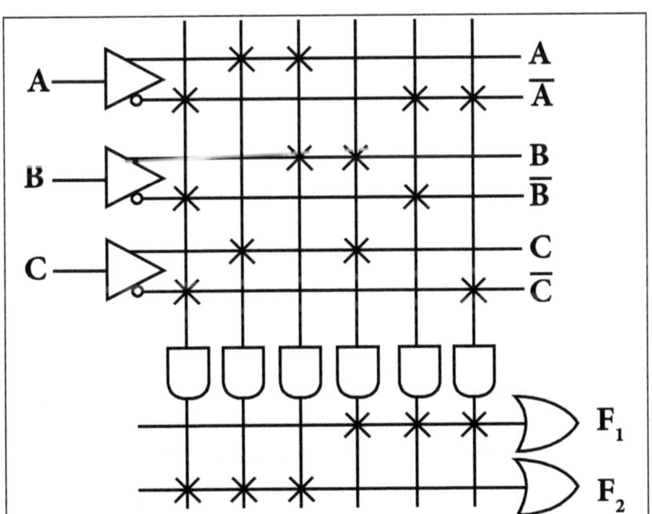

Logic diagram.

4. Let us design an AND - OR PLA that implements the functions:

$$f(x, y, z)=\sum m(0, 2, 4, 6)$$

$$g(x, y, z)=\sum m(1, 3, 5, 7)$$

Solution:

Given:

$$f(x, y, z) = \sum m(0, 2, 4, 6)$$

$$g(x, y, z) = \sum m(1, 3, 5, 7)$$

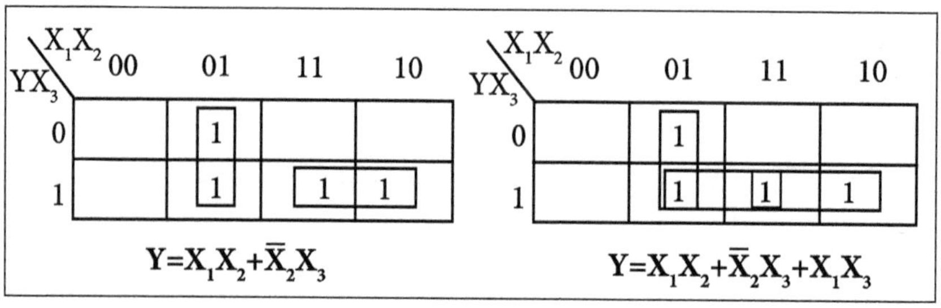

K–Map.

Truth Table

Product Term	Inputs			Outputs		
	x	y	z	f 1 T	G 1 C	T/C
	-	-	0			

Logic diagram:

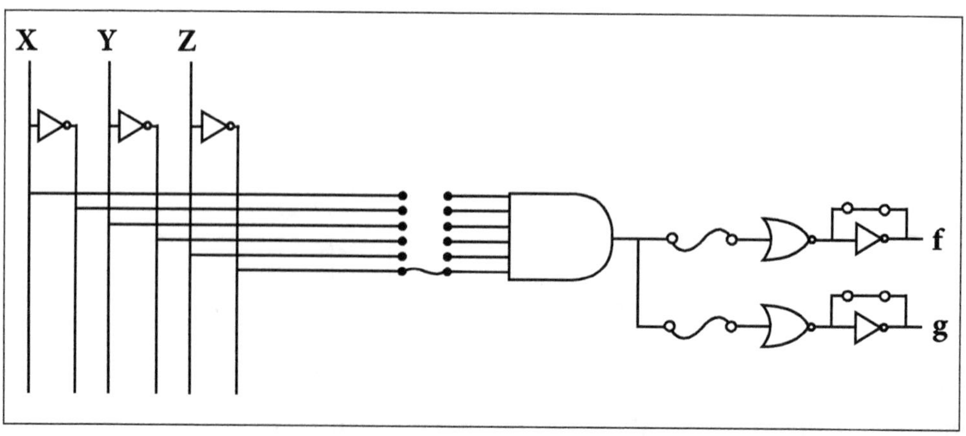

5. A combinational logic circuit which is defined by the following function: $f_1(a, b, c) = \sum(0,1,6,7)$, $f_2\ (a, b,c) = \sum(2,3,5,7)$. Let us implement the circuit with a PAL having three inputs, three product terms and two outputs.

Solution:

Given:

$$f_1(a, b, c) = \sum(0,1,6,7),\ f_2\ (a, b,c) = \sum(2,3,5,7)$$

K-Map

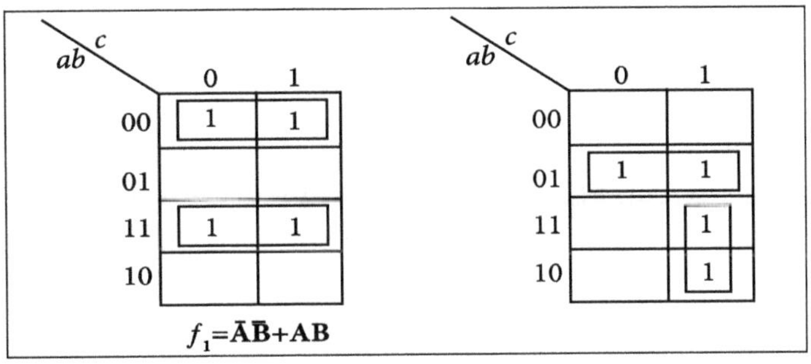

Product	Product	And Inputs			Outputs
Term No	Term	a	b	c	
1		o	o	-	
2	AB	1	1	-	
1	$\bar{A}B$	o	1	-	
2	AC	1	-	1	

Implementation

Logic diagram:

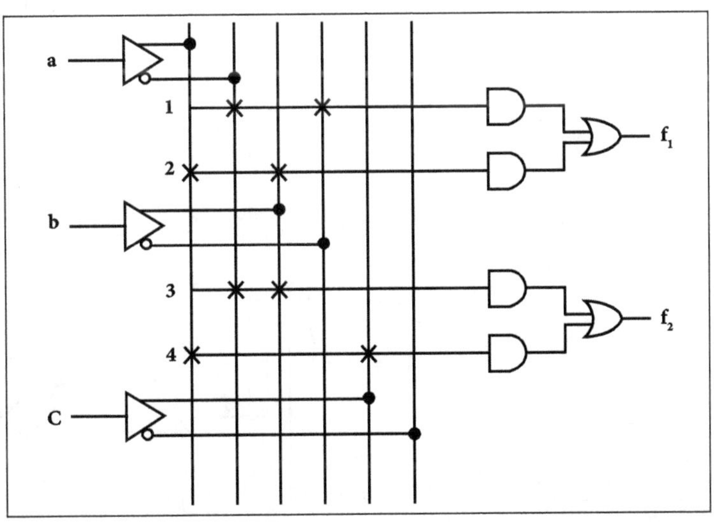

Let us implement a combinational circuit with PLA having 3 inputs, 3 product terms and two outputs $F_1 = \sum(1, 3,5); F_2 = \sum(5, 6, 7)$.

Solution:

Given:

$$F_1 = \sum(1, 3, 5); F_2 = \sum(5, 6, 7).$$

For F_1

A \ BC	$\overline{BC}$	$\bar{B}C$	BC	$B\bar{C}$
$\bar{A}$	0	1	1	0
A	0	1	0	0

For F_2

A \ BC	$\overline{BC}$	$\bar{B}C$	BC	$B\bar{C}$
$\bar{A}$				
A		1	1	1

$$F_1 = \bar{B}C + \bar{A}C$$

$$F_1 = \bar{B}C + \bar{A}C$$

$$\bar{F}_1 = \overline{(\bar{B}C + \bar{A}C)}$$

$$= (B + \bar{C})(A + \bar{C})$$

$$= AB + A\bar{C} + B\bar{C} + \bar{C}\bar{C}$$

$$= AB + A\bar{C} + \underbrace{B\bar{C} + \bar{C}}$$

$$= AB + \underbrace{A\bar{C} + \bar{C}}$$

$$\bar{F}_1 = AB + \bar{C}$$

$$F_2 = AC + AB$$

$$F_2 = AC + AB$$

Logic diagram:

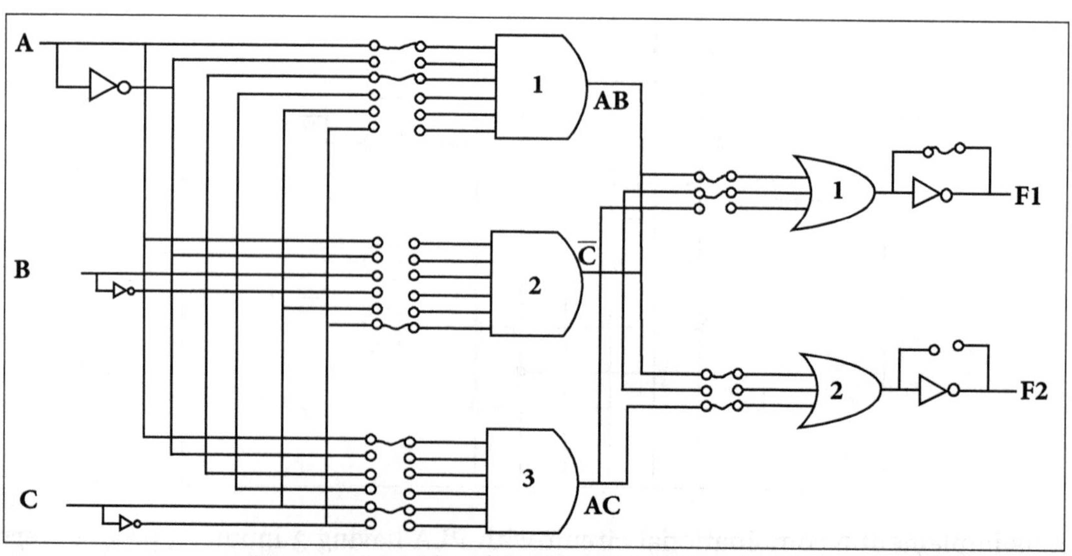

Programming tables of PROM, PAL, PLA Truth table.

Programming Tables of PROM

Example:

Draw a Table using PROM to implement the 2 * 2 bit multiplication.

Serial number	Input to decoder arrays				Output			
	A	B	C	D	F_A	F_B	F_C	F_D
1	0	0	0	0	0	0	0	0
2	0	0	0	1	0	0	0	0
3	0	0	1	0	0	0	0	0
4	0	0	1	1	0	0	0	0
5	0	1	0	0	0	0	0	0
6	0	1	0	1	0	0	0	1
7	0	1	1	0	0	0	1	0
8	0	1	1	1	0	0	1	1
9	1	0	0	0	0	0	0	0
10	1	0	0	1	0	0	1	0
11	1	0	1	0	0	1	0	0
12	1	0	1	1	0	1	1	0
13	1	1	0	0	0	0	0	0
14	1	1	0	1	0	0	1	1
15	1	1	1	0	0	1	1	0
16	1	1	1	1	1	0	0	1

Programming Tables of PLA

PLA program table consists of three columns after the serial number column:

- First column gives the product term in literals.
- Second column gives the inputs in numeric form 0 or I. Cell is filled I when an input variable is present in the product term. Cell is filled to be as 0 when an input variable is present in the product term in complement form.
- Third column gives the AND gate output for a term for each logic function output.

Example:

Draw the PLA program table for code converter functions for the 4-bit BCD output with 4-bit binary number in the input.

Solution:

Given:

The table gives PLA programming table for the Boolean functions. There are both AND and OR programmable links.

Serial number	Product	Inputs					Outputs				
	term	Cy_{00}	A	B	C	D	$F_{CY_{01}}$	FA	FB	FC	FD
1	D	-	-	-	-	1	-	-	-		1
2	$\overline{A} \cdot C$	-	0	-	1	-	-	-	-	1	-
3	$A \cdot B \cdot \overline{C}$	-	1	1	0	-	-	-	-	1	-
4	$\overline{A} \cdot B$	-	0	1	-	-	-	-	1	-	-
5	$B \cdot C$	-	-	1	1	-	-	-	1	-	-
6	$A \cdot \overline{B} \cdot \overline{C}$	-	1	0	0	-	-	1	-	-	-
7	$A \cdot C$	-	1	-	1	-	1	-	-	-	-
8	$A \cdot B$	-	1	-	1	-	1	-	-	-	-
9	Cy_{00}	1	-	-	-	-	1	-	-	-	-

Programming Tables of PAL

PAL program table consists of three columns:

- First column gives the product term number in the SOP.
- Second column gives the required path between the inputs and outputs.
- Third term give the OR gate outputs from the PAL. OR links not fusible. Following example demonstrates how to make the programming table for the PAL.

Example:

Let us draw the program table for code converter functions for the 4-bit BCD output with 4-bit binary number in the input.

Solution:

Given:

The Table gives the programming table for PAL to implement 4-bit binary number to BCD converter. There are only the AND inputs programmable links in PAL programming. The programming table is made as follows:

Serial number	Product term	Inputs					Outputs				
		Cy_{00}	A	B	C	D	$F_{CY_{01}}$	FA	FB	FC	FD
1	D	-	-	-	-	1	$F_{CY_{01}} = D$				
2	$\overline{A}\cdot C$	-	0	-	1	-	$F_c = \overline{A}\cdot C + A\cdot B\cdot \overline{C}$				
3	$A\cdot B\cdot \overline{C}$	-	1	1	0	-					
4	$\overline{A}\cdot B$	-	0	1	-	-	$F_B = \overline{A}\cdot B + B\cdot C$				
5	$B\cdot C$	-	-	1	1	-	$F_A = A\cdot \overline{B}\cdot \overline{C}$				
6	$A\cdot \overline{B}\cdot \overline{C}$	-	1	0	0	-	$Cy_{01} = A\cdot C + A\cdot C + Cy_{00}$				
7	$A\cdot C$	-	1	-	1	-					
8	$A\cdot B$	-	1	-	1	-					
9	Cy_{00}	1	-	-	-	-					

5

Sequential Circuits I

5.1 Classification of Sequential Circuits (Synchronous and Asynchronous)

Combinational circuits and systems create an output based on input variables only. Sequential circuits' uses current input variables and the previous input variables by storing information and putting back into circuit on next clock cycle.

There are 2 types of input to combinational logic circuit. They can be given below as:

- External inputs that come from outside circuit design and are not controlled by circuit.
- Internal inputs which are a function of a previous output states. The internal inputs and outputs are referred to as "secondary's". Secondary inputs are the state variables produced by the storage elements, whereas secondary outputs are excitations for the storage elements.

Two Main Types of Sequential Circuits

The two main types of sequential circuit are given below:

- Synchronous sequential circuits.
- Asynchronous sequential circuits.

Synchronous types use a pulsed or level inputs and a clock input to drive the circuit (with restrictions on pulse width and circuit propagation).

Asynchronous sequential circuits do not use a clock signal as synchronous circuits. Instead the circuit is driven by the pulses of the inputs.

A pulsed output is an output that lasts for the duration of a particular input pulse but can be less in some cases. For the clocked sequential circuits, the output pulse is the same duration as the clock pulse.

A level output refers to an output that changes state at the start of an input pulse or clock pulse and remains in that state until the next input or clock pulse.

5.2 Basic Flip-Flops, Truth Tables and Excitation Tables

In electronics, a flip-flop or latch is the circuit which has two stable states and can be used to store the state information. A flip-flop is a bistable multi vibrator. The circuit can be made to change the state by signals applied to one or more control inputs and that will have one or two outputs. It is an important element of such circuits.

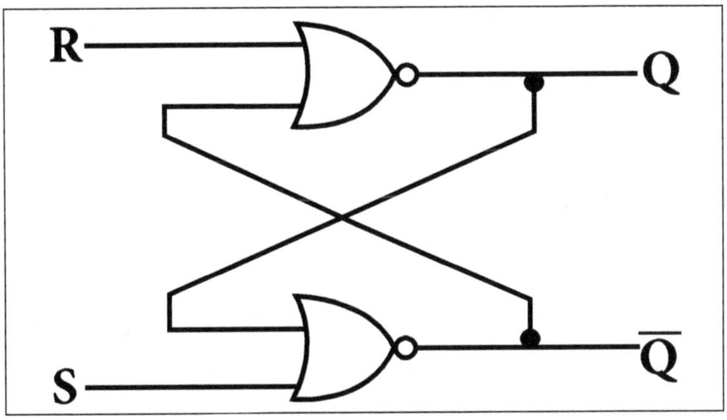

Latch or Flip flop.

The basic unit for storage is flip flop. It maintains its output state either at 1 or 0 till it is directed by an input signal to change its state.

Different Types of Flip-Flop

There are different types of flip flops as follows:

- SR flip-flop
- D flip-flop
- JK flip-flop
- T flip-flop

Sr Flip Flop

The circuit remains in a particular output state indefinitely until something is done to change its output status. If the LOW and HIGH outputs are regarded as 0 and 1 respectively, then the output can either be 0 or 1.

Since either 0 or 1 can be held indefinitely until the circuit is appropriately triggered to go to the other state, the circuit is said to have memory. It is capable of storing one binary digit or one bit of digital information.

Also, in the functioning of the bistable multivibrator circuit, when one of the transistors is in saturation, the other will be in cut-off.

This implies that if we had the outputs from the collectors of both the transistors, then the two outputs would be complementary.

In the flip-flops of various types that are available in IC form, provides complementary outputs usually designated as Q and Q'.

The R-S flip-flop is the basic flip-flop. The letters S and R stands for SET and RESET.

When the flip-flop is SET, its Q output goes to state 1 and when it is RESET it goes to state 0. The Q output is the complement of the Q output at all times.

Operation of RS Flip-Flop

- When the R input is low and the S input is high, the Q output of flip-flop is set.
- When R input is high and S input is low, the Q output of the flip-flop is reset.
- When both the inputs R and S are high, the output is unpredictable.
- When both the inputs R and S are low, the output does not change.

Truth Table

			NAND Gate		
S	R	Q(t + 1)	A	B	Y
0	0	Indeterminate	0	0	1
0	1	Set	0	1	1
1	0	Reset	1	0	1
1	1	NC	1	1	0

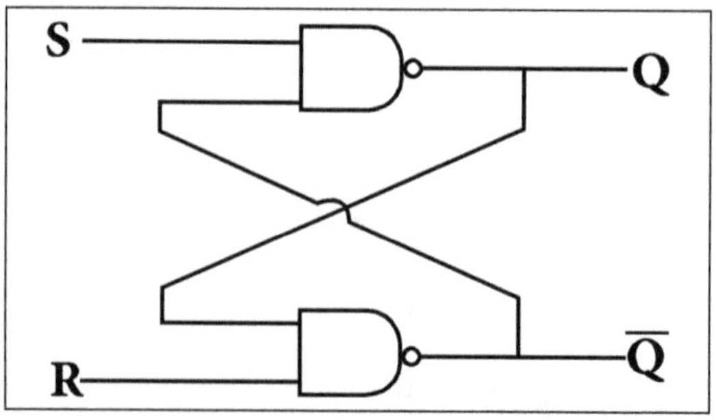

Logic diagram of SR flip-flop.

Latch Flip Flop

The R-S (Reset Set) flip flop is the simplest flip flop of all and easiest to understand. It is basically a device which has two outputs one output being the inverse or complement of the other and two inputs.

A pulse on one of the inputs to take on a particular logical state. The outputs will then remain in this state until a similar pulse is applied to the other input. The two inputs are called the Set and Reset input (sometimes called the preset and clear inputs).

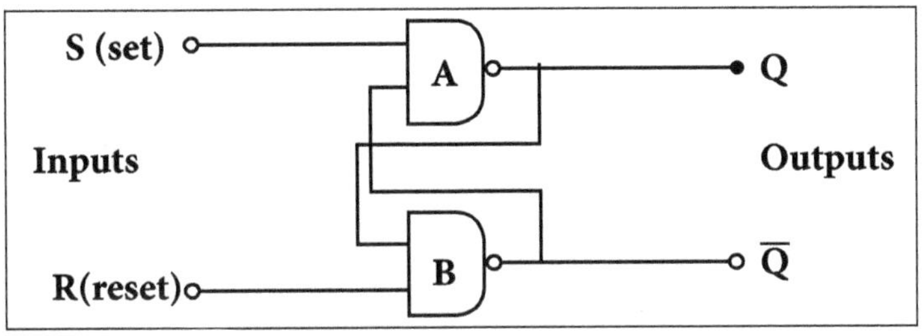

(a) Latch Flip Flop NAND Gate.

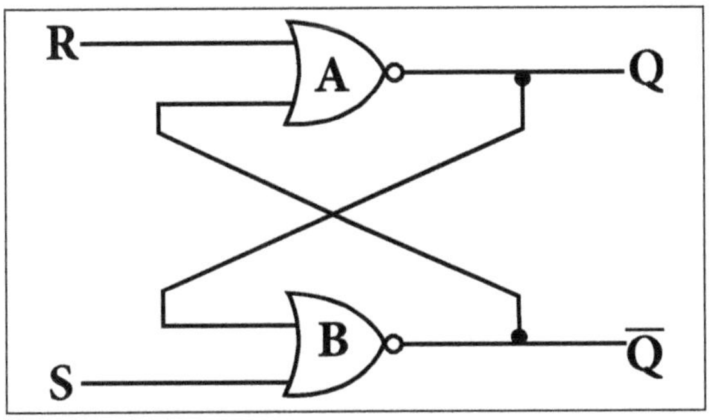

(b) RS Latch Flip Flop NOR Gate.

To describe the circuit of the above figure (a), let us assume that initially both R and S are at the logic 1 state and that output is at the logic 0 states.

Now, if Q = 0 and R = 1, then these are the states of inputs of gatc B, therefore the outputs of gate B is at 1 (making it the inverse of Q i.e., 0). The output of gate B is connected to an input of gate A so if S = 1, both inputs of gate A are at the logic 1 state.

This means that the output of gate A must be 0 (as was originally specified). In other words, the 0 state at Q is continuously disabling gate B so that any change in R has no effect. Also the 1 state at (1 is continuously enabling gate A so that any change S will be transmitted through to Q. The above conditions constitute one of the stable states of the device referred to as the Reset state since Q = 0.

Now, if the R-S flip flop in the Reset state, the S input goes to 0. The output of gate i.e., Q will go to 1 and with Q = 1 and R = 1, the output of gates B (0 will go to 0 with Q now 0 gate A is disabled keeping Q at 1. Consequently, when S returns to the 1 state it has no effect on the flip flop whereas a change in R will cause a change in the output of gate B.

The above conditions constitute the other stable state of the device, called the Set state since Q = 1.The change of the state of S from 1 to 0 has caused the flip flop to change from the Reset state to the Set state.

There is another input condition that is, when both the R and S inputs are taken to the logic state 0. When this happens both Q and Q will be forced to 1 and will remain so far as long as R and S are kept at 0.

However when both inputs return to 1 there is no way of knowing whether the flip flop will latch in the Reset state or the set state. The condition is said to be indeterminate because of this indeterminate state great care must be taken when using R-S flip flop to ensure that both inputs are not instructed simultaneously.

Simple NAND R-S Flip Flop Truth table is described in the following table:

S	R	Q
0	0	indeterminate
0	1	Set(1)
1	0	Reset(0)
1	1	No change

The Truth table for the NAND R-S flip flop is given below:

Initial conditions	Inputs(Pulsed)		Final output	
Q	S	R	Q	
1	0	0	indeterminate	
1	0	1	1	0
1	1	0	0	1
1	1	1	1	0
0	0	0	indeterminate	
0	0	1	1	0
0	1	0	0	1
0	1	1	0	1

When NOR gate are used the R and S inputs are transposed compared with the NAND version. Also the stable state when R and S are both 0. A change of state is effected by pulsing the appropriate input to the 1 state. The indeterminate state is now when both R and S are simultaneously at logic 1.

NOR Gate R-S Flip Flop Truth Table

S	R	Q
0	0	No change
0	1	Reset(0)
1	0	Set(1)
1	1	Indeterminate

Clocked RS Flip Flop

The RS latch flip flop required the direct input but no clock. It is very useful to add clock to control precisely the time at which the flip flop changes the state of its output.

In the clocked R-S flip flop the appropriate levels applied to their inputs are blocked till the receipt of a pulse from another source called clock. The flip flop changes state only when clock pulse is applied depending upon the inputs.

The basic circuit is shown in the below figure. This circuit is formed by adding two AND gates at inputs to the R-S flip flop. In addition to control inputs Set (S) and Reset (R), there is a clock input (C) also.

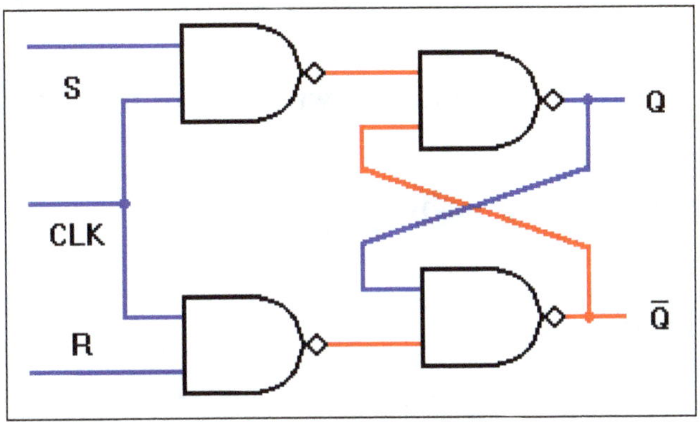

The Truth Table for the Clocked R-S Flip Flop

Initial conditions	Inputs(Pulsed)		Final output	Comment
Q	S	R	Q(t + 1)	No change
0	0	0	0	No change
0	0	1	0	Clear Q
0	1	0	1	Set Q
0	1	1	???	indeterminate
1	0	0	1	No change
1	0	1	0	Clear Q
1	1	0	1	Set Q
1	1	1	???	indeterminate

Excitation Table for R-S Flip Flop

S	R	Q
0	0	No change
0	1	Reset(0)
1	0	Set(1)
1	1	Indeterminate

JK Flip Flop

The J-K flip-flop behaves in the same fashion as an R-S flip-flop except for one entry in the function table.

In case of R-S flip-flop, its input combination S = R = 1 and the input combination S = R = 0 are prohibited.

In the case of the J-K flip-flop with active HIGH inputs, the output of the flip-flop toggles, (i.e.,) it goes to the other state, for J = K = 1.

The output toggles for J = K = 0 in thte case of the flip-flop having a active LOW input. Thus, the J-K flip-flop can overcome the problem of a forbidden input combination of the R-S flip-flop.

The corresponding Karnaugh maps are shown in below figure for the characteristics table.

The characteristic equations is given by:

$$Q_{n+1} = J.\overline{Q_n} + \bar{K}.Q_n$$

$$Q_{n+1} = \bar{J}.\overline{Q_n} + K.Q_n$$

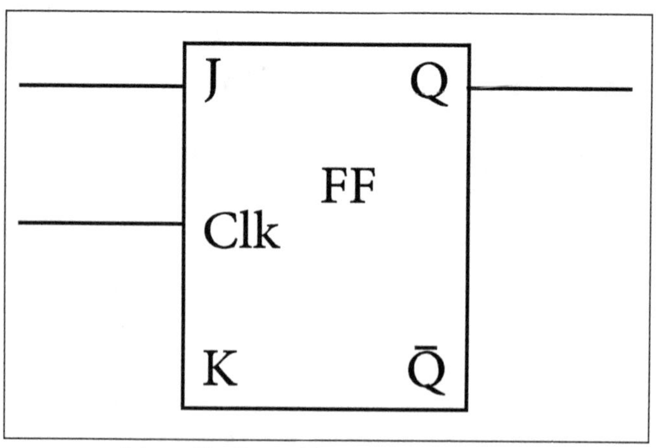

(a) JK flip flop with active high inputs.

Truth Table

Operation mode	J	K	Clk	Qn+1
SET	1	0	1	1
RESET	0	1	1	0
NO CHANGE	0	0	1	Qn
TOGGLE	1	1	1	$\overline{Q}_n$

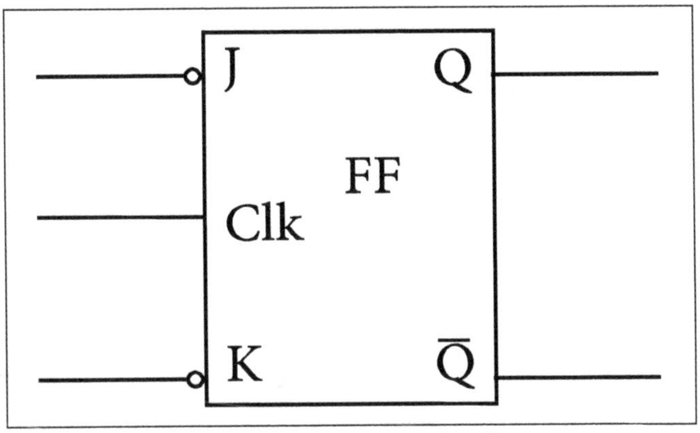

(b) JK flip flop with active low inputs.

Truth Table

Operation mode	J	K	Clk	Qn+1
SET	0	1	1	1
RESET	1	0	1	0
NO CHANGE	1	1	1	Qn
TOGGLE	0	0	1	$\bar{Q}_n$

JK-Flip Flop has the following features:

- When J = 1 and K = I, the output of SIC flip flops toggles (changes to opposite state) on a clock edge.
- When J= 0 and K= 0, the output of JK flip flops does not change on a clock edge.
- When J= 0 and K = I, the output Q resets to 0 after the clock edge.
- When J= I and K =0, the output Q sets to I after the clock edge.

Truth Table

Input					Output		State
J	$\bar{Q}$	CLK_1	K	Q	Q_{n+1}	$\bar{Q}_{n+1}$	$(n+1)^{th}$
1	0	1 at the + ve edge	1	1	0	1	Toggle $Q_{n+1} = \bar{Q}_n$
1	1	1 at the + ve edge	1	0	1	0	Toggle $Q_{n+1} = \bar{Q}_n$
1	0	-ve edge or 0 or 1	1	1	1	0	Q_n No Change
1	1	-ve edge or 0 or 1	1	1	1	0	Q_n No Change

JK Flip-Flop is a Universal Flip-Flop

When configured in various ways, JK flip flop is capable of operating like most other types of flip - flops.

Operation of JK flip-flop:

- When the K input is low and the J input is high, the Q output of flip-flop is set.
- When the K input is high and the J input is low, the Q output of flip-flop is reset.
- When both the inputs K and J are low, the output does not change.
- When both the inputs K and J are high, it is possible to set or reset the flip-flop (i.e.,) the output toggle on the next positive clock edge.

State Diagram

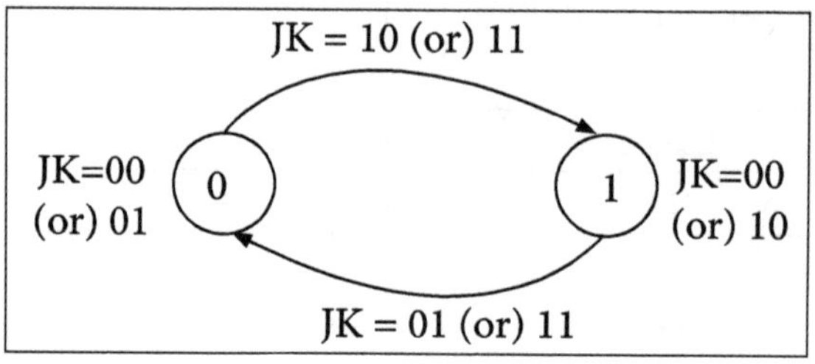

State Table

Q	J	K	Q(t + 1)
0	0	0	0
0	0	1	0
0	1	0	1
0	1	1	1
1	0	0	1
1	0	1	0
1	1	0	1
1	1	1	0

Counter using JK Flip Flop

Q_2	Q_1	Q_0
0	0	0
0	0	1
0	1	1
1	1	1
1	1	0

1	0	0
0	0	0

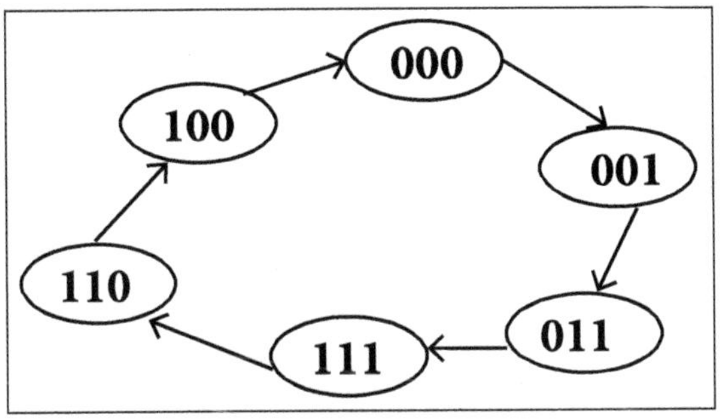

State Diagram.

State Table

PS			NS			JK Input					
Q2	Q1	Q0	Q0 + 1	Q2 + 1	Q1 + 1	J2	K2	J1	K1	J0	K0
0	0	0	1	0	0	0	X	0	X	1	X
0	0	1	1	0	1	0	X	1	X	X	0
0	1	0	0	0	0	0	X	X	1	0	X
0	1	1	1	1	1	1	X	X	0	X	0
1	0	0	0	0	0	X	1	0	X	0	X
1	0	1	0	0	0	X	1	0	X	X	1
1	1	0	0	1	0	X	0	X	1	0	X
1	1	1	0	1	1	X	0	X	0	X	1

K - Map

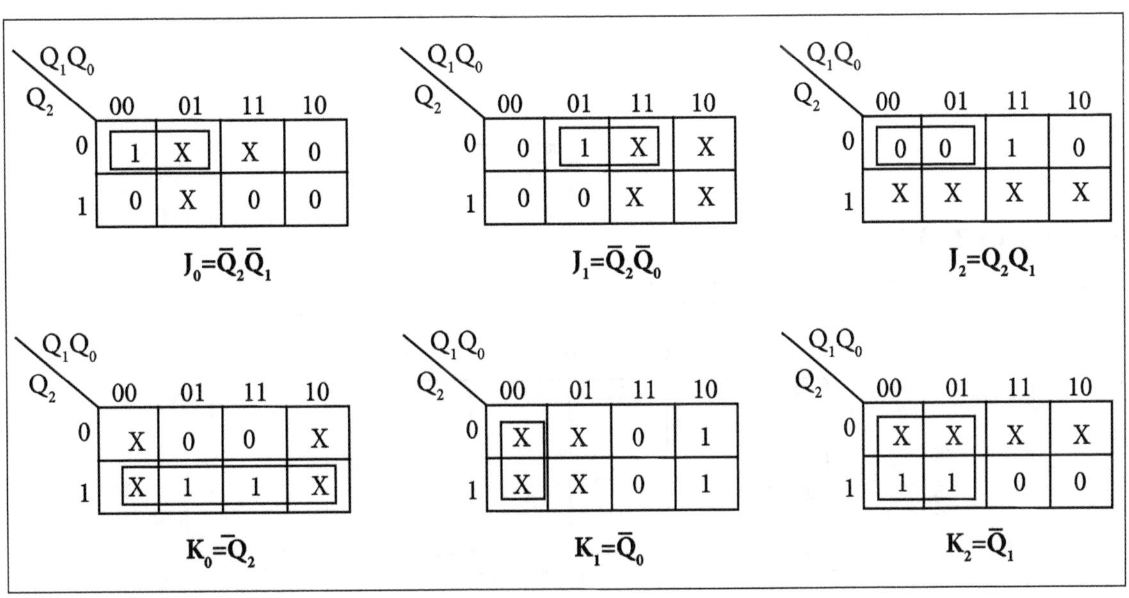

Logic Diagram:

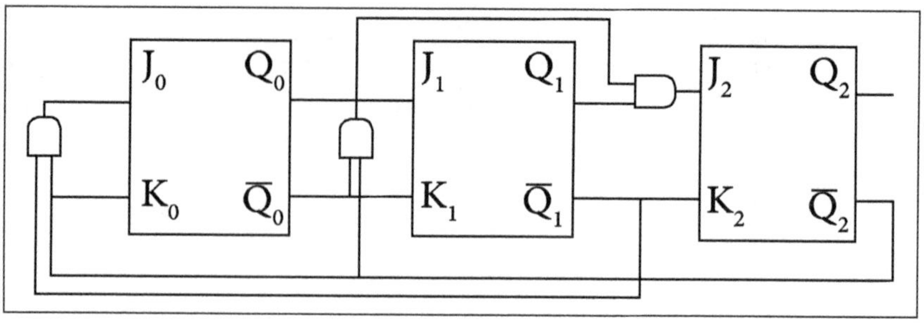

D Flip-Flop

A D flip-flop also called as a delay flip-flop can be used to provide temporary storage of one bit of information.

The below figure shows the symbol of the circuit and the function table of a negative edge-triggered D flip-flop.

When the clock is active, the data bit (0 or 1) present at the D input is transferred to the output.

In the D flip-flop, the data transfer from D input to Q output occurs on the negative-going transition of the clock input. D input can acquire new status.

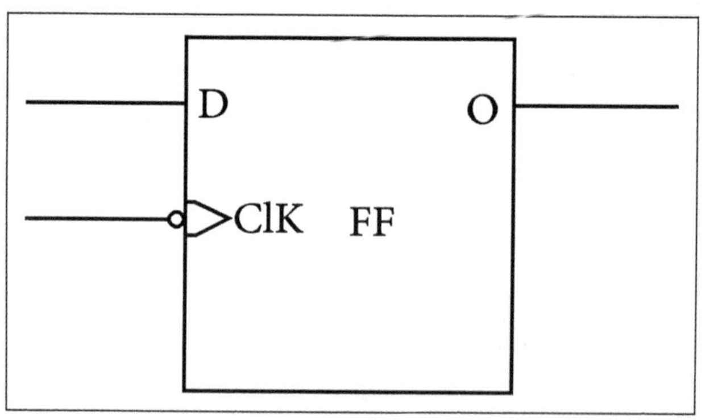

Symbol

Function Table

D	ClK	Q
0	↓	0
1	↓	1

Truth Table

Q_n	D	Q_{n+1}
0	0	0
0	1	1
1	0	0
1	1	1

K-Map

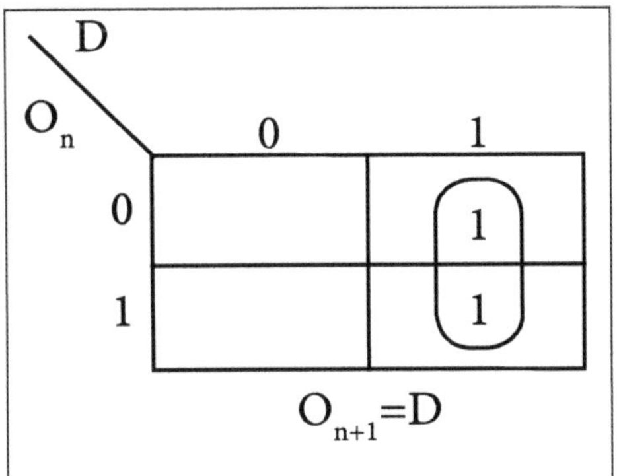

(d)

Operation of D Flip-flop

In D flip-flop during the occurrence of clock pulse if D=1, the output Q is set and if D=0, the output is reset.

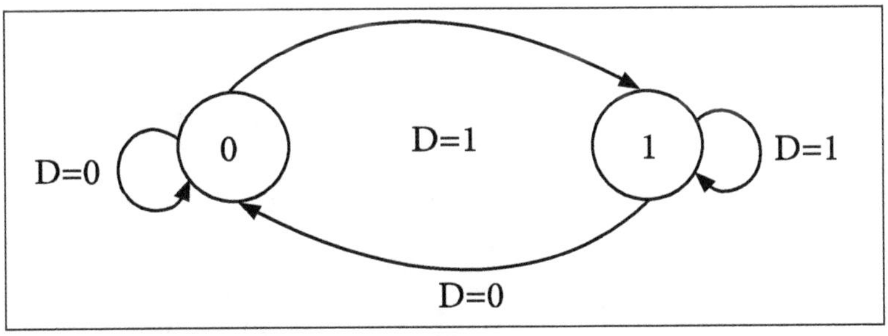

State diagram.

State Table

Q	D	Q(t + 1)
0	0	0
0	1	1
1	0	0
1	1	1

Flip-Flop Excitation Tables for D Flip-Flop

In D flip-flop, its next state is always equal to the D input and it is independent of the present state.

D is 0 if Q_{n+1} has to 0 and if Q_{n+1} has to be 1 regardless the value of Q_n.

T Flip-Flop

The output of a toggle flip-flop also called a T flip-flop changes state every time it is triggered at its T input known as the toggle input. That is, its output becomes 1, if it was 0 and 0 if it was 1.

The characteristic equations can be written as follows:

$$Q_{n+1} = T.\overline{Q_n} + \overline{T}.Q_n$$
$$Q_{n+1} = \overline{T}.\overline{Q_n} + T.Q_n$$

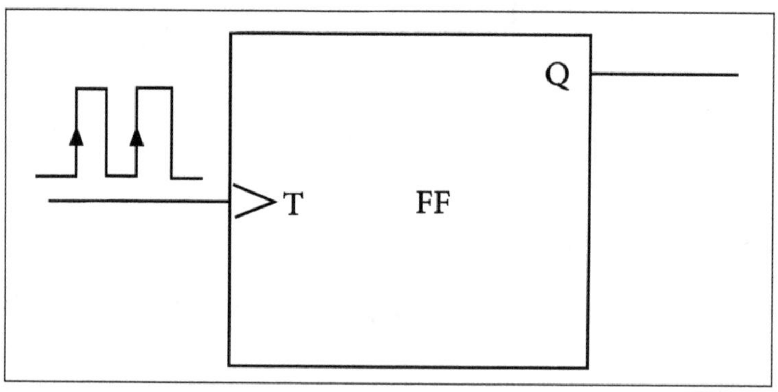

T flip-flop

T	Q_n	Q_{n+1}
↑	0	1
↑	1	0

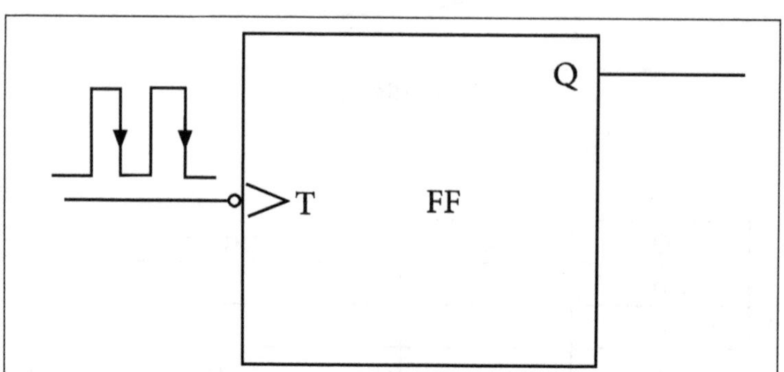

T flip-flop

T	Q_n	Q_{n+1}
↓	0	1
↓	1	0

Truth Table

Q_n	T	Q_{n+1}
0	0	0
0	1	1
1	0	1
1	1	0

Operation of T flip-flop

T flip-flop is also known as Toggle flip-flop.

When the value of T = 0, there is no change in the output.

When the value of T = 1, the output switch to the complement state (i.e.,) the output toggles.

Flip-Flop Excitation Tables for T Flip-Flop

When input T = 1, the state of the flip-flop is complemented.

When T = 0, the state of the flip-flop remains unchanged. Therefore, for 00 and 11 transitions T must be 0 and for 01 and 10 transitions T must be 1.

5.3 Conversion from One Flip-Flop to another Flip-Flop

JK Flip Flop to SR Flip Flop

The S and R will be the external inputs to J and K. As shown in the logic diagram below, the J and K values will be the outputs of the combinational circuit. Hence, the values of J and K have to be obtained in terms of S, R and Q_p.

The conversion table is to be written using S, R, Q_p, Q_{p+1}, J and K. For two inputs S and R, eight combinations can be made. For each combination, its corresponding Q_{p+1} outputs are found out.

The outputs for the combinations of S = 1 and R = 1 are not permitted for an SR flip flop. Thus, the outputs are considered as invalid and the J and K values are taken as "don't cares".

J-K Flip Flop to S-R Flip Flop

Conversion Table

S-R Inputs		Outputs		J-K Inputs	
S	R	Qp	Qp+1	J	K
0	0	0	0	0	X
0	0	1	1	X	0
0	1	0	0	0	X
0	1	1	0	X	1
1	0	0	1	1	X
1	0	1	1	X	0
1	1	Invalid		Dont care	
1	1	Invalid		Dont care	

Logic Diagram

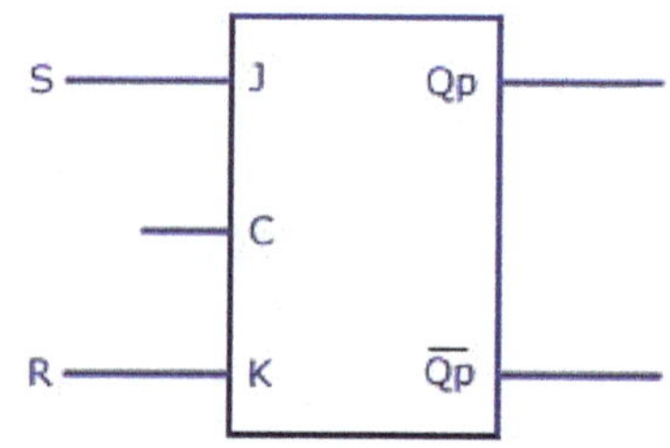

K-maps

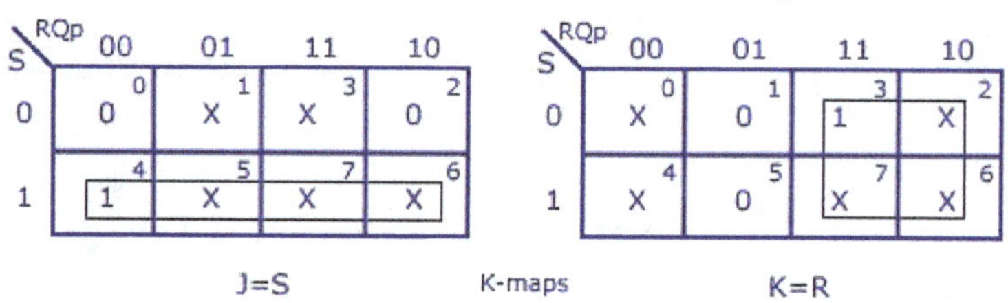

Conversion table, K-Map and Logic diagram.

SR Flip Flop to D Flip Flop

The S and R are the actual inputs of the flip flop and D is the external input of the flip flop as shown in the below figure The four combinations, the logic diagram, conversion table and the K-map for S and R in terms of D and Q_p are shown below.

S-R Flip Flop to D Flip Flop

Conversion Table

D Input	Outputs		S-R Inputs	
	Qp	Qp+1	S	R
0	0	0	0	X
0	1	0	0	1
1	0	1	1	0
1	1	1	X	0

K-maps

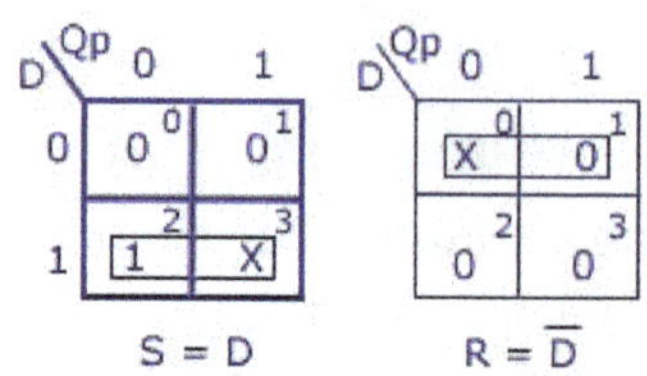

Logic Diagram

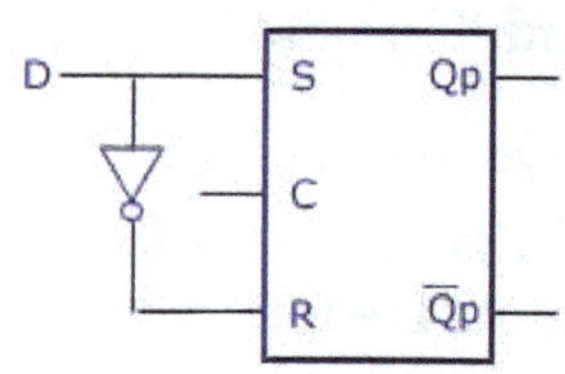

Conversion table, K-Map and Logic diagram.

D Flip Flop to SR Flip Flop

The D is the actual input of the flip flop and S and R are the external inputs. Eight possible combinations are achieved from the external inputs S, R and Q_p. As the combination of S=1 and R=1 are invalid, the values of Q_{p+1} and D are considered as "don't cares".

The logic diagram showing the conversion from D to SR and the K-map for D in terms of S, R and Q_p are shown below:

D Flip Flop to S-R Flip Flop

Conversion Table

S-R Inputs		Outputs		D Input
S	R	Qp	Qp+1	
0	0	0	0	0
0	0	1	1	1
0	1	0	0	0
0	1	1	0	0
1	0	0	1	1
1	0	1	1	1
1	1	Invalid		Dont care
1	1	Invalid		Dont care

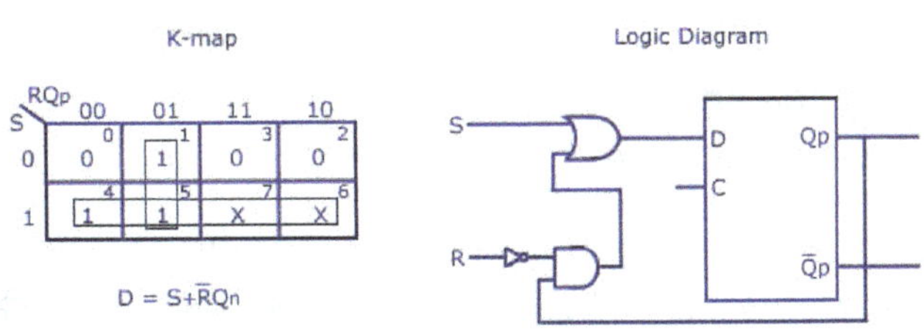

Conversion table, K-Map and Logic diagram.

JK Flip Flop to T Flip Flop

The J and K are the actual inputs of the flip flop and T is taken as the external input for conversion. Four combinations are produced with T and Q_p. The J and K are expressed in terms of T and Q_p.

J-K Flip Flop to T Flip Flop

Conversion Table

T Input	Outputs		J-K Inputs	
	Qp	Qp+1	J	K
0	0	0	0	X
0	1	1	X	0
1	0	1	1	X
1	1	0	X	1

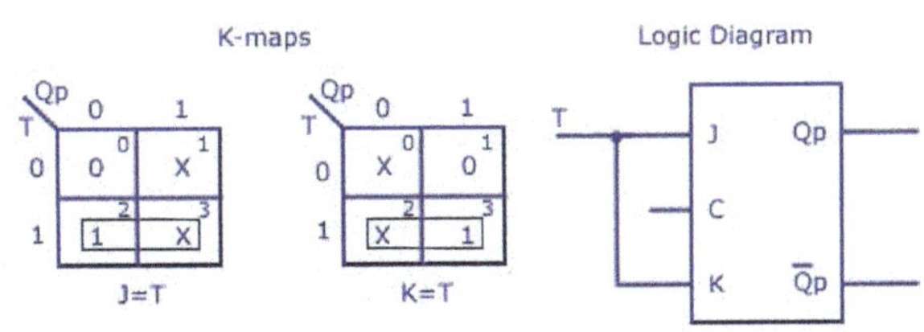

Conversion table, K-Map and Logic diagram.

5.4 Design of Ripple Counters

A ripple counter is an asynchronous counter where only the first flip-flop is been clocked by an external clock. Every subsequent flip flops are clocked by the output of the preceding flip-flop. The asynchronous counters are also known as ripple-counters since the way in which the clock pulse ripples it through the flip-flops.

The MOD of the asynchronous counter or ripple counter is 2^n if n flip-flops are used. For a 4-bit counter, the range of the count is 0000 to 1111 (2^4-1). The counter may count up or count down or count up and down depending on the input control.

Thus, the count sequence usually repeats itself. During the process of counting up, the count sequence shall go from 0000, 0001, 0010, ... 1110, 1111, 0000, 0001, ... etc., whereas when counting down the count sequence shall go in the opposite manner: 1111, 1110, ... 0010, 0001, 0000, 1111, 1110, ... etc.

Thus, the complement of the count sequence counts in reverse direction. If the un-complemented output counts up, the complemented output counts down and vice versa.

There are many ways to implement the ripple counter depending on the characteristics of the flip flops used and requirements of the count sequence.

The clock trigger is either positive edged or negative edged. The flip flops used are JK or D flip-flops. The count direction takes place is up, down or up/down.

The asynchronous counters are slower than synchronous counters because of the delay in transmission of the pulses from flip-flop to flip-flop. In synchronous circuit, all the bits in the counter changes synchronously with an assertion of the clock. The examples of synchronous counters are the Ring and Johnson counter.

This can be implemented using D-type flip-flops or JK-type flip-flops.

The below figure uses 2 D flip-flops to implement a divide-by-4 ripple counter ($2^n = 2^2 = 4$). It counts down.

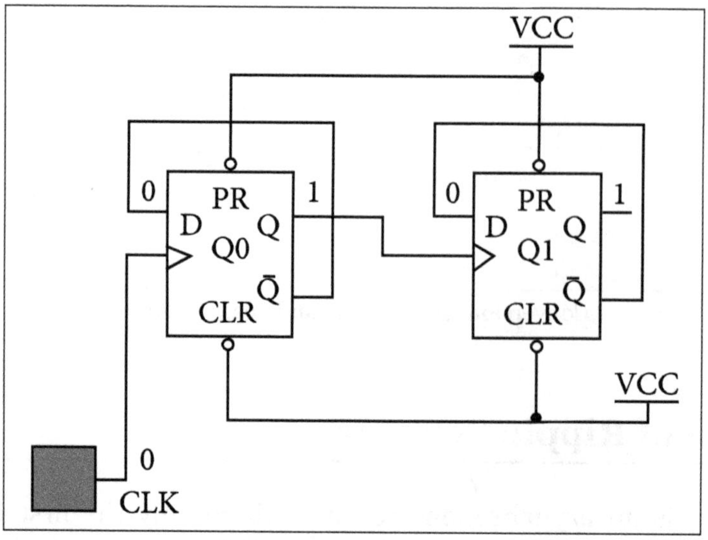

Divide-by-4 ripple counter.

State Table

Count	Q1	Qo
3	1	1
2	1	0
1	0	1
0	0	0

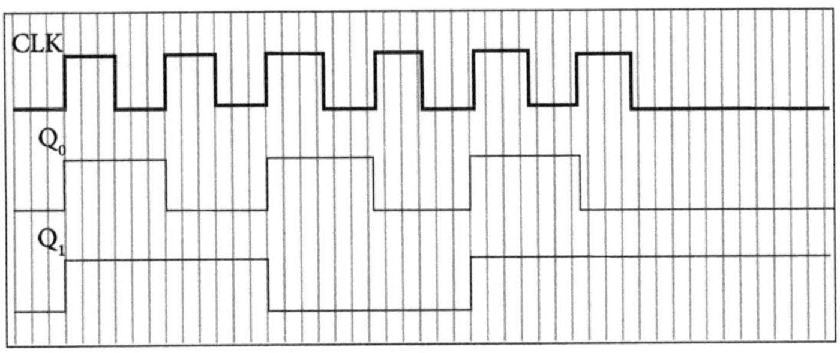

Timing diagram.

5.5 Design of Synchronous Counters: Johnson Counter and Ring Counter

Procedure to Design a Synchronous Counter

- Step 1: Obtain the truth table of the logic sequence for intended counter to be designed. Alternatively obtain the state diagram of the counter.
- Step 2: Determine number and type of flip-flop to be used.
- Step 3: From the excitation table, determine its next state logic.
- Step 4: From the output state, use Karnaugh map for simplification to derive the circuit output functions and flip-flop output functions.
- Step 5: Draw the logic circuit diagram.
- Step 6: Simulate the circuit using software.
- Step 7: Build the circuit.

The steps to design a Synchronous Counter using JK flip flops are:

The general sequential circuit in terms of its basic parts and its input and outputs.

Let us design a 2 bit up/down counter with an input D which determines the up/down function. Thus when D = 0, the count sequence is 00, 01, 10, 11, 00... When D = 1, the count sequence is 00, 11, 10, 01, 00...

Solution:

Step 1: Draw the state diagram for the given sequence.

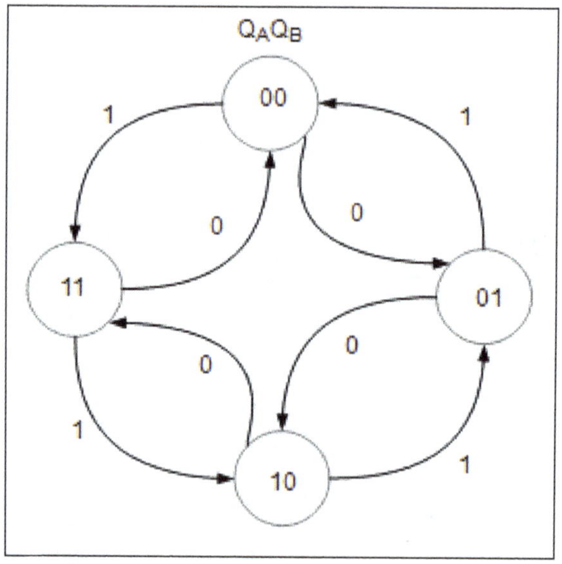

State diagram.

Step 2: Develop a next-state table for the specific counter sequence by using the state diagram as a reference and fill up the present state and next state columns. For this interactive table, we can modify the next state.

Present State			Next State		JK flip flop inputs			
D	QA	QB	QA	QB	JA	KA	JB	KB
0	0	0	0	1	0	X	1	X
0	0	1	1	0	1	X	X	1
0	1	0	1	1	X	0	1	X
0	1	1	0	0	X	1	X	1
1	0	0	1	1	1	X	1	X
1	0	1	0	0	0	X	X	1
1	1	0	0	1	X	1	1	X
1	1	1	1	0	X	0	X	1

JK Flip Flop Truth Table

J	K	Q
0	0	Qo
0	1	0
1	0	1
1	1	Toggle

JK Flip Flop Transition Table

Q_N	Q_{N+1}	J	K
0	0	0	X
0	1	1	X
1	0	X	1
1	1	X	0

Step 3: Use K-map to derive the logic equations.

JA = D QB + D QB

QAQB \ D	0	1
00	0	1
01	1	0
11	X	X
10	X	X

KA = D QB + D QB

QAQB \ D	0	1
00	X	X
01	X	X
11	1	0
10	0	1

JB = 1

QAQB \ D	0	1
00	1	1
01	X	X
11	X	X
10	1	1

KB = 1

QAQB \ D	0	1
00	X	X
01	1	1
11	1	1
10	X	X

Step 4: Use the Boolean expressions to implement the counter.

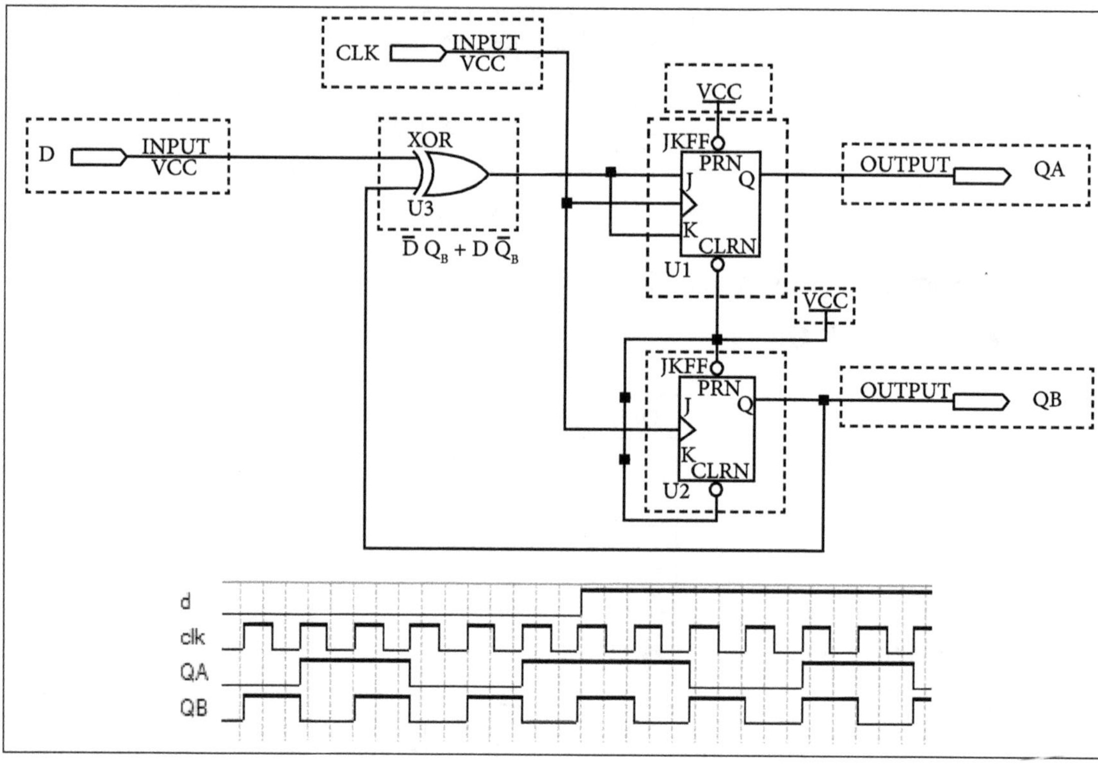

5.5.1 Johnson Counter and Ring Counter

Johnson Counter

A Johnson counter is a K-bit switch-tail ring counter with the 2k decoding gates for providing the outputs for the 2k timing signals. The Johnson counter is said to be a special case of shift register where, the output from the last stage is being inverted and fed back as input to the first stage.

The pattern of bits is equal in length to the shift register so it circulates indefinitely. These counters are sometimes also called as "walking ring" counters. It is used in the specialist applications including those which are similar to the decade counter, digital to analogue conversion, etc.

Ring Counter

The ring counter is a circular shift register which has only one flip flop that is being set at any particular time, all others are cleared. It is a counter which counts up and when it reaches the last number which is designed to count up, it will reset itself back to the first number.

Example: The ring counter which is designed using 3 JK Flip Flops may count starting from 001 to 010, 010 to 100 and back to 001. It will repeat itself in a 'Ring'. Therefore, it is called as a ring counter.

5.6 Design of Registers

A clocked sequential circuit consists of a group of flip-flops and combinational gates connected to form a feedback path. The flip-flops are essential because in their absence, the circuit reduces to a purely combinational circuit.

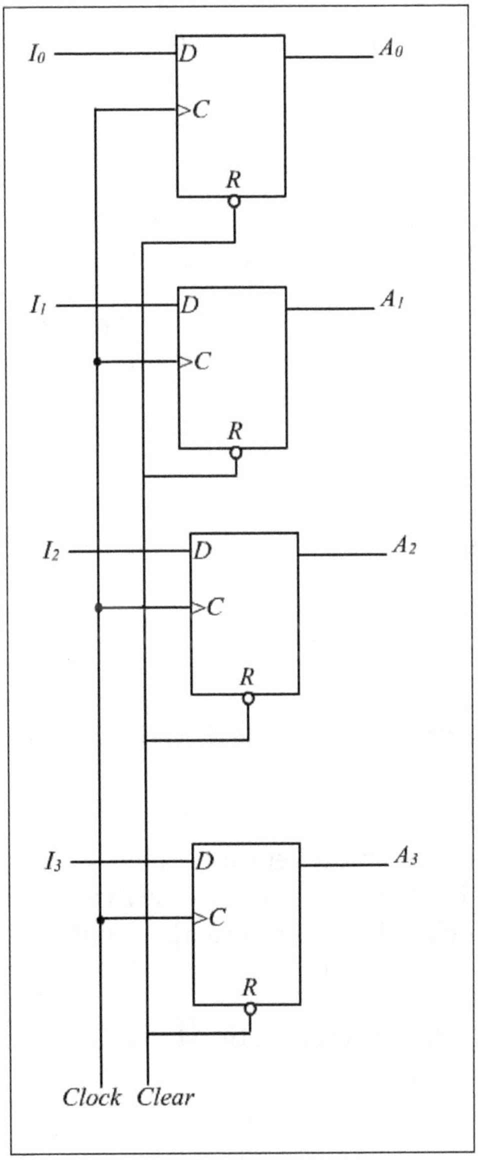

4-bit register.

The circuit with flip-flops is considered as sequential circuit even in the absence of the combinational gates. The circuits that include flip-flops are usually classified by the function they perform rather than by name of the sequential circuit. For example, registers and counters.

A register is a group of flip-flops which is capable of storing one bit of information. A n-bit register consists of a group of n flip-flops capable of storing n bits of binary information. In addition to the flip-flops, register may have combinational gates that perform certain data-processing tasks. The flip-flops hold the binary information and the gates determine how the information is transferred into the register.

The counter must be a register which goes through the predetermined sequence of binary states. Gates in the counter are connected in such a way as to produce the prescribed sequence of states. Although counters are the special type of register, it is common to differentiate them by giving them a different name.

Various types of registers are available commercially. The simplest register is one that consists of only flip-flops, without any gates. The figure shows such a register constructed with four D-type flip-flops to form a 4-bit data storage register.

The common clock input triggers all the flip-flops on the positive edge of each pulse and the binary data available at the four inputs are transferred into the register. The four outputs could be sampled at any time to obtain the binary information stored in the register.

The input Clear_b goes to the active-low R (reset) input of all the four flip flops. When this input goes to 0, all flip-flops are reset asynchronously. The Clear_b input is useful for clearing the register to all 0's prior to its clocked operation. Depending on the type of flip-flop used either Clear, Clear_b, reset or reset_b can be used to indicate the transfer of register to an all 0's state.

5.6.1 Buffer Register and Controlled Buffer Register

Buffer Register

The figure below shows the simplest register which is constructed with four D flip-flops. This register is also known as buffer register. Every D flip-flop is triggered with a common negative edge clock pulse. Therefore, the input X bits set up the flip-flops for loading.

When the first negative clock edge arrives, the stored binary information becomes:

$$Q_{AQBQCQD = ABCD}$$

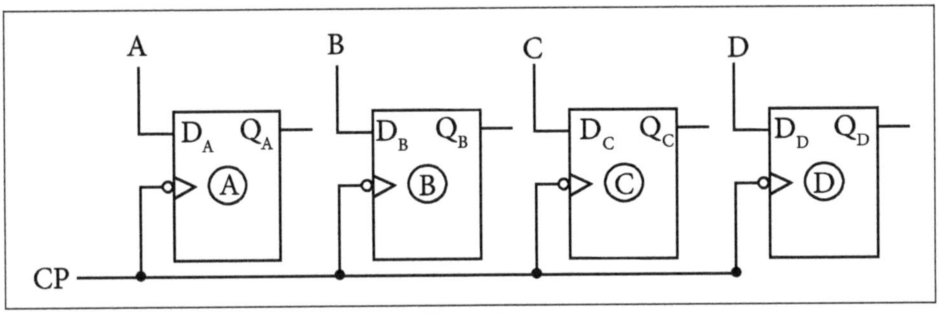

Buffer register.

In buffer register, only D flip-flops are used. Thus, it can store 4-bit binary information. So the number of flip-flop stages in a register will determines its total storage capacity.

Controlled Buffer Register

The input and output of the register can be controlled by connecting the tristate devices at the input and output sides of register as shown in the figure below. Hence this register is known as 'controlled buffer register'.

In controlled buffer register, the tristate switches are used to control the operation. Whenever we want to store data in the register, we can make LOAD or WR signal low in order to activate the tristate buffers.

Similarly, if we want the data at the output, we have to make RD signal low to activate the buffers.

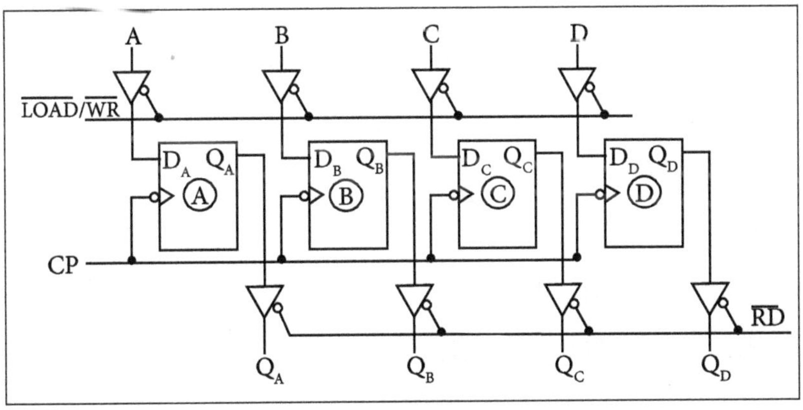

Controlled Buffer Register.

Therefore, the controlled buffer registers are commonly used for temporary storage of data within a digital system.

5.6.2 Shift Register

The register which are capable of shifting its binary information in one or both the directions from a state to a state within the register or into or out of the register upon the application of clock pulses is known as a shift register. This register is capable of shifting the data, one bit at a time. It is used for converting between the serial and parallel interfaces.

It is useful for as many as circuits to work on groups of bits in parallel, but comparatively the serial interfaces are simpler to construct. The shift registers are used as simple delay circuits and it can also be used as pulse extenders. The serial shift register consists of a chain of flip-flops which is connected in cascade with the output of one flip-flop being connected to the input of its neighbor. The basic operation of the shift register is synchronous.

In shift registers, every flip-flop is connected to a common clock. The D flip-flops forms the simplest type of shift-registers. This basic data movements is possible only within a four-bit shift register which is shown in the below figure:

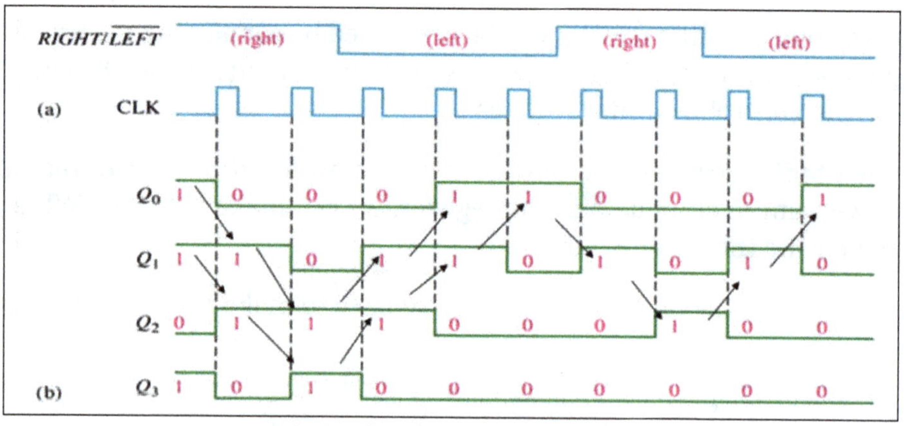

Timing diagram of shift register.

The different types of shift registers are as follows:

- Serial In, Serial Out (SISO).
- Serial In, Parallel Out (SIPO).
- Parallel In, Serial Out (PISO).
- Parallel In, Parallel Out (PIPO).

Serial-In, Serial-Out

Destructive Readouts: In destructive readout, every datum once lost has been shifted out of the right-most bit. It is the simplest kind of shift register. A data string is presented at 'Data In' and it is shifted in right one at stage where each time the 'Data Advance' is brought high. At every advance, the bit on the far left which is 'Data In' that is shifted into the first flip-flop's output. The bit on the far right is 'Data Out' that is shifted out and lost.

Non-Destructive Readout: The non-destructive readout can be achieved, when another input line is added known as the Read/Write Control. If it is high that is write, then the shift register behaves as normal by advancing the input data one place for each clock cycle and the data can be lost from the end of the register.

Since, when the R/W control is set low that is Read, any data can be shifted out of the register at the right becomes the next input at the left and it is kept in the system. Hence, as long as the R/W control is set low, none of the data can be lost from the system.

Serial-In, Parallel-Out

The Serial-in, Parallel-out configuration allows the conversion from serial to parallel format. The data are given as an input serially and once the data has been input, it may be either read off at each output simultaneously or it can also be shifted out and replaced.

Parallel-In, Serial-Out

The Parallel-In, Serial-Out configuration has the data input in the parallel format. For writing the data to the register, the Write/Shift control line must be held LOW. For shifting the data, the W/S control line is brought HIGH and the registers are clocked. Until the number of clock cycles is not more than the length of the data-string, the Data Output, Q shall be the parallel data read off in order.

Parallel-In, Parallel-Out

In this configuration, the shift register takes the data from the parallel inputs (D_0-D_3) and shifts it to the corresponding output (Q_0-Q_3) when the registers are clocked. This can be used as a kind of 'history' for retaining the old information as the input in another part of the system as long as it is ready for new information, where upon the registers are clocked and the new data are 'let through'.

5.6.3 Bi-directional Shift Register

The bidirectional shift register allows the shifting of data either to the left or to the right side. This can be implemented by using the logic gate circuitry which enables the transfer of data from one stage to the next stage where either to the right or to the left, depending on the level of a control line. The figure below shows a four-bit bidirectional register.

The RIGHT or LEFT is the control input signal which allows the data shifting either towards right or towards left. If high on this line, it enables the shifting of data towards the right and a low enables it towards the left. When the RIGHT or LEFT signal is high, the gates G_1, G_2, G_3, G_4 are enabled. The state of the Q output of each flip-flop is passed through the D input of the following flip-flop.

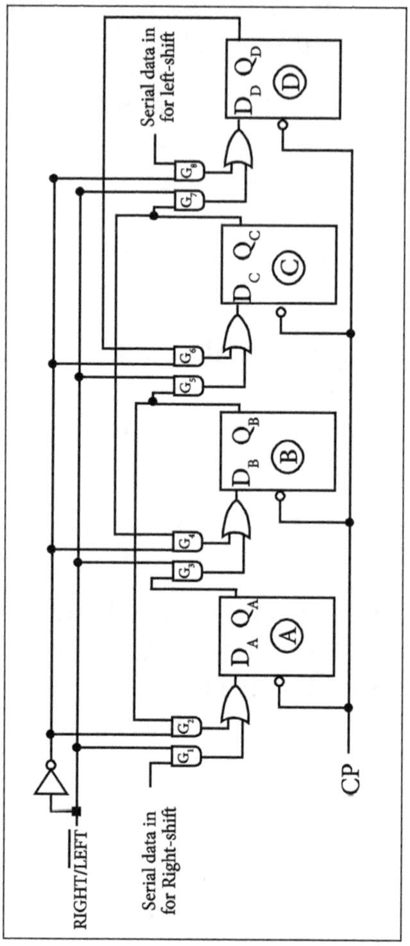

(a) 4-bit bi-directional shift register.

During the arrival of clock pulse, the data are shifted from one place to the right. If the RIGHT or LEFT signal is low, the gates G_5, G_6, G_7, G_8 are enabled. The Q output of each flip-flop is passed through the D input of the preceding flip-flop. During the arrival of clock pulse, the data are shifted from one place to the left.

Bidirectional Shift Register with the Parallel Load

We can see that shift register can be used for converting the serial data into parallel data and vice-versa. While the parallel load capability is added to the shift register, the data entered in parallel can be taken out in serial way by shifting the data stored in the register. This register is known as the Bidirectional Shift Register with parallel load. Figure below shows the bidirectional shift register with parallel load:

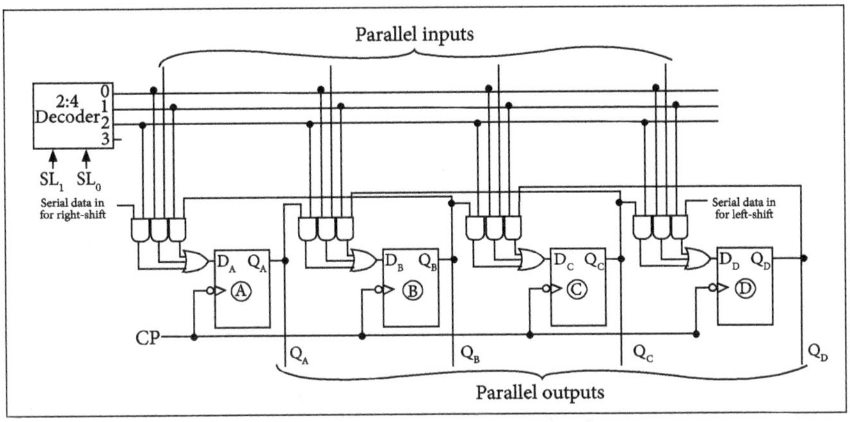

(b) 4-bit bi-directional shift register with parallel load.

As shown in the above figure (b), the D input of each flip-flop has three sources as follows:

- Output of left adjacent flip-flop.
- Output of right adjacent flip-flop.
- Parallel input.

From these three sources, one of the sources is selected at a time and it is done with the help of the decoder. The decoder selects the lines (SL_1 and SL_0) by selecting the one source out of three as shown in the table below:

SL_1	SL_0	Selected Source
0	0	Parallel input
0	1	Output of right adjacent FF
1	0	Output of left adjacent FF
1	1	-

If the select lines are 00 (i.e., SL_1 = 0 and SL_0 = 0), the data from the parallel inputs is loaded into the 4-bit register. If the select lines are 01 (i.e., SL_1 = 0 and SL_0 = 1), the data within the register is shifted to 1-bit left. If the select lines are 10 (i.e., SL_1 = 1 and SL_0 = 0), the data within the register is shifted to 1-bit right.

5.6.4 Universal Shift Register

There are many high speed bi-directional "universal" type Shift Registers which are namely, TTL 74LS194, 74LS195 or the CMOS 4035 that are available as 4-bit multi-function devices which can be used in either the serial-to-serial, right shifting, left shifting, parallel-to-serial, serial-to-parallel or as a parallel-to-parallel multi-function data register. Thus, it is known as the "Universal".

The universal shift registers may perform any combination of parallel and serial input to output operations but they require additional inputs to specify the desired function and to pre-load and reset the device. A commonly used universal shift register is the TTL 74LS194.

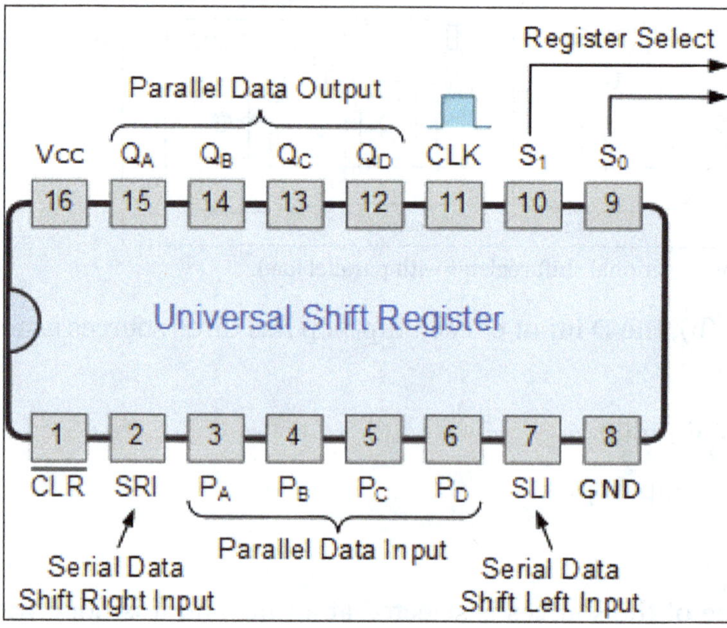

Mode Control		Register Type
S_1	S_0	
0	0	Hold
0	1	Shift to right
1	0	Shift to left
1	1	Parallel Mode

4-bit Universal Shift Register 74LS194.

The universal shift registers are very useful digital devices. This can be configured to respond to the operations that require some form of temporary memory storage or to transfer the data from one point to another in either a serial or parallel format or for the delay of information such as the SISO or PIPO configuration modes.

The universal shift registers are commonly used in the arithmetic operations for shifting data to the left or right for multiplication or division.

6

Sequential Circuits II

6.1 Finite State Machine and Analysis of Clocked Sequential Circuits

A finite-state machine or FSM is a model of computation based on a hypothetical machine made of one or more states. Only a single state can be active at the same time, so the machine must transit from one state to another in order to perform different actions. FSM's are commonly used to organize and represent an execution flow, which is useful to implement AI in games.

Analysis of Clocked Sequential Circuits

The analysis of a synchronous sequential circuit is the process of determining the functional relation that exists between its outputs and its internal states. The contents of all the flip-flops in the circuit combined will determine the internal state of the circuit. Thus, if the circuit contains n flops, it can be in one of the r states.

Knowing the present state of the circuit and the input values at any time t, we should be able to derive its next state (i.e., the state at time t+1) and the output produced by the circuit at t. A sequential circuit can be described completely by a state table that is very similar to the ones shown for flip-flops. For a circuit with n flip-flops, there will be 2^n rows in the state table.

If there are m inputs to the circuit, there will be 2^m columns in the state table. At the intersection of each row and column, the next-state and the output information are recorded. A state diagram is a graphical representation of the state table, in which each state is represented by a circle and the state transitions are represented by arrows between the circles. The input combination that brings about the transition and the corresponding output information are shown in the below figure.

Sequential Circuit Analysis

Analyzing a sequential circuit thus corresponds to generate the state table or the state diagram for the circuit. The state table or state diagram can be used to determine the output sequence generated by the circuit for a ten input sequence if the initial state is known.

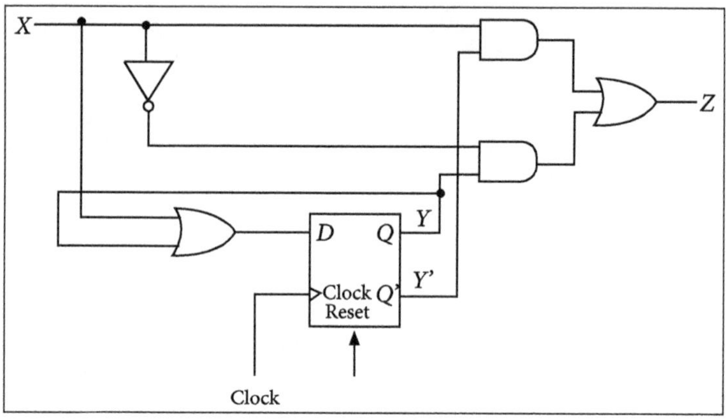

(a) Circuit

Q(r + 1)

Q \ X	0	1
0	0	1
1	1	1

Z

Q \ X	0	1
0	0	1
1	1	0

(b) next-state and output tables.

Q \ X	0	1
0	0/0	1/1
1	1/1	1/0

(c) transition table.

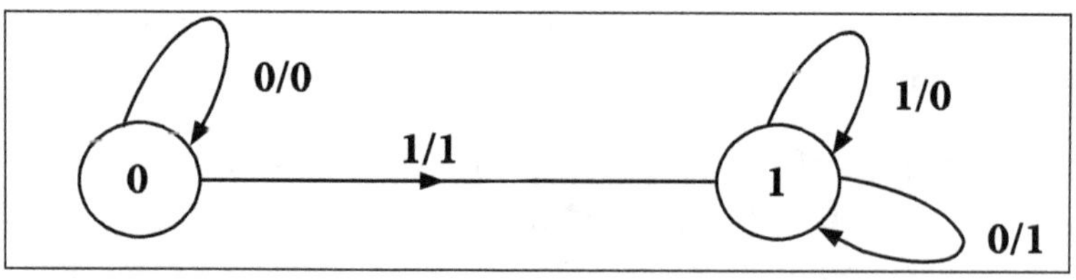

(d) State diagram.

It is important to note that for proper operation, a sequential circuit must be in its initial state before the inputs applied. Usually the power-up circuits are used to initialize the circuit to the appropriate state when the power is turned on.

Analysis

It consists of a table or a diagram for the time sequence of inputs, outputs and interval states. A logic diagram is recognized as a synchronous sequential circuit if it includes flip - flops. The flip - flops may be of any type and the logic diagram may or may not include combinational circuit gates.

Design

It consists of choosing the flip flops and combinational gate structure together with the flip - flops. The number of flip-flop is determined from the number of states needed in the circuit. The combinational circuit is derived from the state table.

6.2 State Diagrams and State Tables

A state represents the status of the flip flops in a sequential circuit. If there are n flip flops then the circuit can have 2^n possible states.

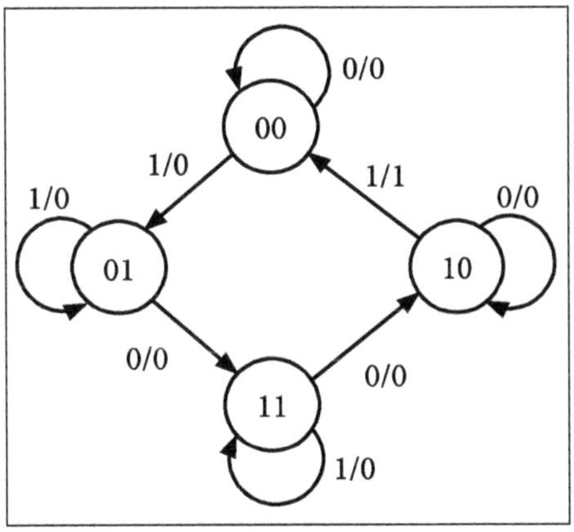

State diagram.

State diagram is the graphical representation of the information available in a state table. In state diagram, a state is represented by a circle and the transitions between states are indicated by directed lines connecting the circles. In addition to graphical symbols, tables or equations, flip-flops can also be represented graphically by a state diagram.

In the above mentioned diagram, a state is represented by a circle and the transition between states is indicated by the directed lines (or arcs) connecting the circles. An example of a state diagram is shown in the above figure.

State Table

State table representation of a sequential circuit consists of three sections labeled as present state, next state and output. The present state designates the state of flip-flops before occurrence of a clock pulse.

The next state shows the states of flip-flops after the clock pulse and the output section lists the value of the output variables during the present state. A state table gives the time sequence of inputs, outputs and flip-flop states.

6.3 Reduction of State Tables, State Assignment and Design Procedures

The problem of state reduction is to find ways of reducing the number of states in a sequential circuit without altering the input-output relationships. In other words, to reduce the number of states, redundant states should be eliminated. A redundant state Si is a state which is equivalent to another state S_j.

Two states are said to be equivalent if, for each member of the set of inputs, they give exactly the same output and send the circuit either to the same state or to an equivalent state.

Since 'm' flip-flops can describe a state machine of up to 2^m states, reducing the number of states may (or may not) result in a reduction in the number of flip-flops.

For example, if the number of states is reduced from 8 to 5, we still need 3 flip-flops.

However, state reduction will result in more don't care states. The increased number of don't care states can help obtain a simplified circuit for the state machine.

Let us consider the below state diagram:

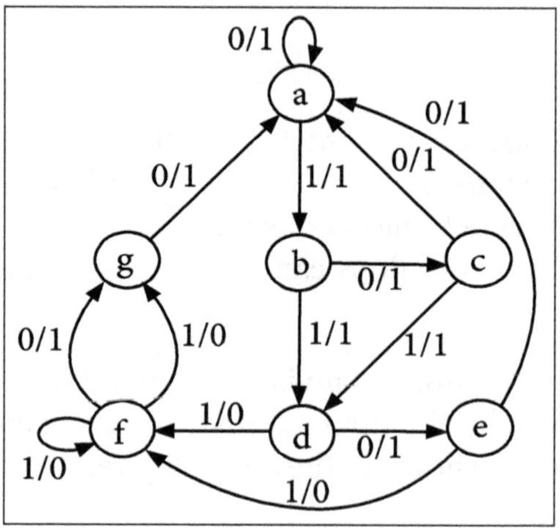

State Diagram.

The state reduction proceeds by first tabulating the information of the state diagram into its equivalent state table form (as shown in the below table).

The problem of state reduction requires identifying equivalent states. Each N state is replaced by 1 state.

Let us consider the following state table.

States 'g' and 'e' produce the same outputs, i.e., '1' and '0' and take the state machine to same next-states, 'a' and 'f', on inputs '0' and '1' respectively. Thus, states 'g' and 'e' are equivalent states.

We can now remove state 'g' and replace it with 'e' as shown in the below state table.

We next note that the above change has caused the states 'd' and 'f' to be equivalent. Thus in the next step, we remove state 'f' and replace it with 'd'.

There are no more equivalent states remaining. The reduced state table results in the following reduced state diagram:

Present state	Next state		Output	
	X=0	X=1	X=0	X=1
a	a	b	1	1
b	c	d	1	1
c	a	d	1	1
d	o	~~f~~ d	1	0
e	a	~~f~~ d	1	0
~~f~~	~~g~~ e	f	1	0
~~g~~	a	f	1	0

State Assignment

State assignment is the process of assigning binary values to states in such a way as to reduce the cost of the combinational circuit that drives the FFs. This is particularly useful when a sequential circuit is reviewed from its external input-output terminal. Binary values of the individual states are of no consequence as long as a circuit produces the required sequence of outputs for a number of given inputs.

- The cost of the combinational circuit of a sequential circuit can be reduced by using the known simplification methods for combinational circuits.
- State assignment procedures are concerned with methods for assigning binary values to states in such a way as to reduce the cost of combinational circuit that drives the flip - flops.
- After the number of states in a state table has been reduced, the next step in realizing the transition table (state-assigned table) is to assign flip-flop states (i.e., binary values) to correspond to the states in the state table.
- The cost of the logic required to realize a sequential circuit is strongly dependent on the way this state assignment is made.

Techniques used for State Assignment

- Shared row state assignment.
- One hot state assignment.

There are three possible ways of binary assignment can be done as follows:

State	Assignment -1			Assignment -2			Assignment -3		
a	0	0	1	0	0	0	0	0	0
b	0	1	0	0	1	0	1	0	0
c	0	1	1	0	1	1	0	1	0
d	1	0	0	1	0	0	1	0	1
e	1	0	1	1	1	1	0	1	0

Assignment-1 is straight binary sequence (a to e). The other two are chosen arbitrarily. So different binary assignments result in a state table with different binary values for the states although the input-output relationship remains the same.

However, this binary form of the state table can easily be used to derive the combinational circuit part of the sequential circuit. However, to date, there are no state assignment procedures that guarantee a minimal cost combinational circuit.

After getting the minimum state diagram as well as the state table, the next step is that the designer should choose secondary variables properly to locate the variable states. The required numbers of secondary variables are determined by the total number of states in the reduced state diagram. Assume a state diagram as depicted in the below state diagram which has four distinct states. Therefore, two secondary variables are required to define each state uniquely.

The number of secondary variables which are used to define all states of the reduced state diagram is equal to the number of flip-flops required to implement the sequential circuit. As there are four states and two secondary variables, A and B required to describe all four states, only two flip-flops are required to implement the circuit. The state assignment for the state S_0 is AB = 00. Similarly, the state assignment for states S_1, S_2 and S_3 are AB = 01, AB = 10 and AB = 11 respectively.

State	Assignment
S_0	00
S_1	01
S_2	10
S_3	11

State assignment for S_0, S_1, S_2 and S_3.

The cost of the combinational logic circuit of a sequential circuit can be reduced by a state assignment properly. If the minimized or reduced state diagram of a sequential machine has five states S_0, S_1, S_2, S_3, S_4 the possible three combinations of state assignment are illustrated in the below table. The assignment-1 is a straight binary for the sequence of states S_0, S_1, S_2, S_3, S_4. The assignment-2 and 3 are chosen arbitrarily. The sequential circuit can be implemented any one of these assignments but the designer should choose the state assignment correctly so that the cost of the circuit is minimum.

State	Assignment-1	Assignment-2	Assignment-3
So	000	010	000
S1	001	000	011
S2	010	011	001
S3	011	001	100
S4	100	100	010

State assignment for S_0, S_1, S_2, S_3 and S_4.

Procedures for Designing Sequential Circuits

Design of a sequential circuit that can be carried out, using state-diagrams and state-tables.

The design procedure is as follows:

- The word description of the behavior is stated. This is done by the state-diagram, a timing diagram or other permanent information.
- From the given information obtain the state-table.
- Reduce the number of states, by state-reduction method, if the sequential circuit can be characterized by input-output relationships independent of the number of states.
- Assign binary values to each state, if contain letter symbols.
- Deduce the number of FFs needed and assign a letter symbol to each.
- Chose the type of FF to be used.
- From the state table, derive the circuit-excitation and output tables.
- Using map or any other simplification method derives the circuit-output function and FE UP function.
- Draw the logic diagram.

Sequential Logic Circuits

The two states are said to be equivalent if, for each member of the set of inputs, they give exactly the same output and send the circuit to the same state or to an equivalent state.

In certain cases the states are specified in letter symbols. In such cases there comes another factor, called state assignment. State assignment procedures are concerned with methods for assigning binary values to states in such a way as to reduce the cost of the combinational circuit that drives the flip-flop.

For any problem there may be a number of different state assignments leading to different combinational parts of the sequential circuit. The most common criterion is that the chosen assignment should result in a simple combinational circuit for the flip-flop inputs. However, to date, there are no state assignment procedures that guarantee a minimal-cost combinational circuit.

Design of the clocked sequential circuit whose state diagram is given below:

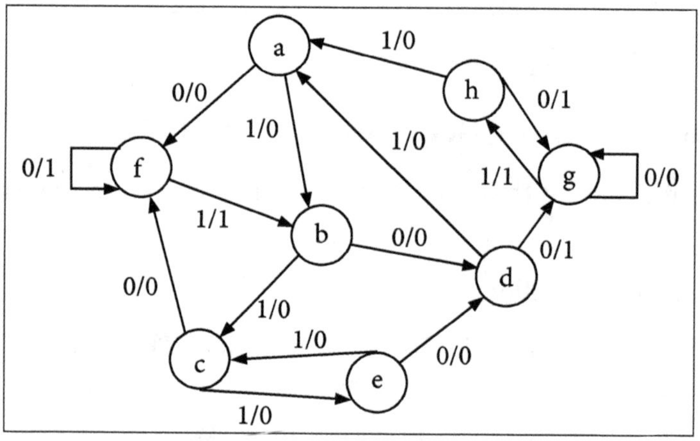

The state table for the above state diagram is shown below:

Present state	Next state		Output	
	X = 0	X = 1	X = 0	X = 1
a	f	b	0	0
b	d	c	0	0
c	f	e	0	0
d	g	a	1	0
e	d	c	0	0
f	f	b	1	1
g	g	h	0	1
h	g	a	1	0

- We now look for two equivalent states, find that d and h are two such states. They both go to g & a and have outputs of 1 and 0 for x = 0 & x = 1, respectively. Therefore, states d and h are equivalent; one can be removed.
- Similarly, we find that b and e are again two such states. They both go to d and c and have outputs of 0 and 0 for x = 0 and x = 1, respectively.

- Therefore, states b and e are also equivalent and one can be removed. The procedure of removing a state and replacing it by its equivalent is demonstrated in the table which is given below.

- From the below table we find that present state c now has next states f and b and outputs 0 and 0 for x = 0 and x = 1, respectively. The same next states and outputs appear in the row with present state a. Therefore, states a and c are equivalent, state c can be removed and replaced by a.

Present state	Next state		Output	
	X = 0	X = 1	X = 0	X = 1
a	f	b	0	0
b	d	~~c~~ a	0	0
~~c~~	f	~~e~~ b	0	0
d	g	A	1	0
~~e~~	d	C	0	0
F	f	b	1	1
g	G	~~h~~ d	0	1
~~h~~	G	a	1	0

The final reduced state table is shown in the below table:

Present state	Next state		Output	
	X = 0	X = 1	X = 0	X = 1
a	f	b	0	0
b	d	a	0	0
d	g	a	1	0
f	f	b	1	1
g	g	d	0	1

The state diagram for the reduced state table consists of only five states and is shown in the below figure:

We now assign the different states the binary values. As there may be a variety of state assignments. Some of them are shown in the below table. Among them we may choose any of them and accordingly let us design the circuit.

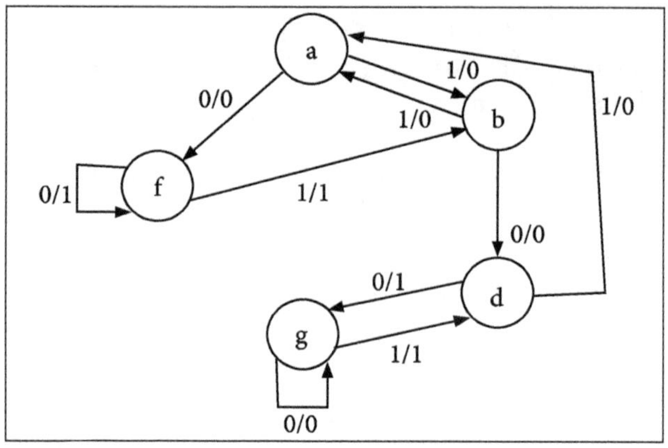

State	Assignment1	Assignment2	Assignment3	Assignment4
a	000	001	111	011
b	001	010	001	101
d	010	011	110	111
f	011	100	101	001
g	100	101	010	000

The below figure contains the state table, we have used binary assignment 1 to substitute the letter symbols of the five states. It is obvious that a different binary assignment will result in a state table, with completely new binary values for the states while the input-output relationships will remain the same.

Present state	Next state		Output	
	X = 0	X = 1	X = 0	X = 1
000	011	001	0	0
001	010	000	0	0
010	100	000	1	0
011	011	001	1	1
100	100	010	0	1

The derivation of the excitation table is facilitated if we arrange the state table in a different form. This form is shown in the below table, where the present state and the input variables are arranged in the form of a truth table. We may use any flip-flop, but the simplest form of the circuit is possible with J-K flip-flops. So we now design the circuit using J-K flip-flops.

Present state			Input	Next state			Flip-flop inputs						Output
A	B	C	x	A	B	C	JA	KA	JB	KB	JC	KC	y
0	0	0	0	0	1	1	0	X	1	X	1	X	0
0	0	0	1	0	0	1	0	X	0	X	1	X	0
0	0	1	0	0	1	0	0	X	1	X	X	1	0

0	0	1	1	0	0	0	0	X	0	X	X	1	0
0	1	0	0	1	0	0	1	X	X	1	0	X	1
0	1	0	1	0	0	0	0	X	X	1	0	X	0
0	1	1	0	0	1	1	0	X	X	0	X	0	1
0	1	1	1	0	0	1	0	X	X	1	X	0	1
1	0	0	0	1	0	0	X	0	0	X	0	X	0
1	0	0	1	0	1	0	X	1	1	X	0	X	1

There are three unused states in this circuit, binary states 101, 110 and 111. When an input of 0 or 1 is included with these unused states, we obtain six don't-care terms. These six binary combinations are not listed in the table under the present state or input and are treated as don't-care terms.

In below figure, Karnaugh maps for JA, KA, JB, KB, JC and KC are shown:

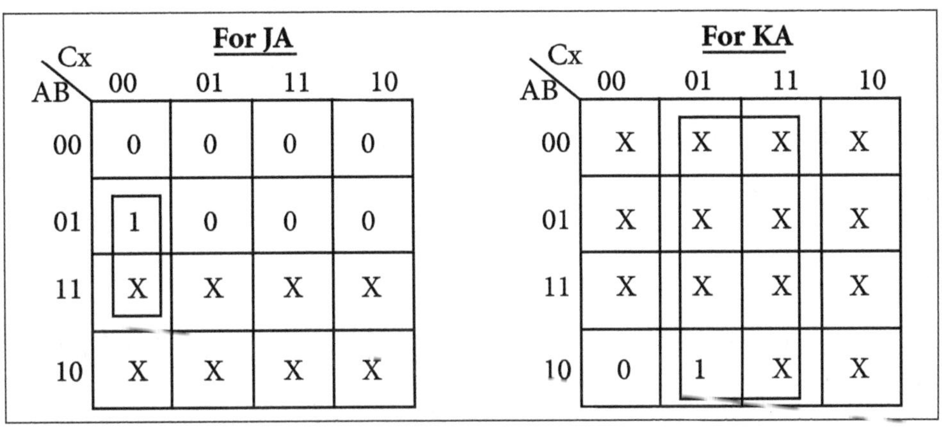

From the Karnaugh maps for JA and KA, we obtain as:

JA=BC'x' and

KA=x.

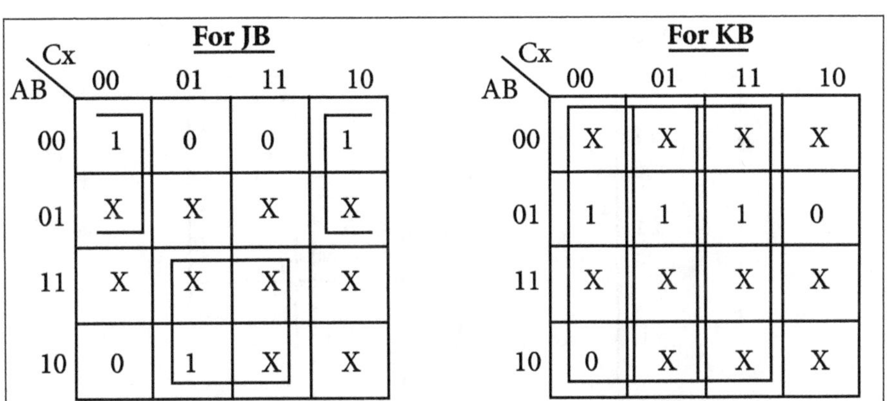

The Boolean expressions are derived for JB and KB from the Karnaugh maps as:

$JB = Ax + A'x'$ and

$KB = C' + x.$

For JC

AB \ Cx	00	01	11	10
00	1	1	X	X
01	0	0	X	X
11	X	X	X	X
10	0	0	X	X

For KC

AB \ Cx	00	01	11	10
00	X	X	1	1
01	X	X	0	0
11	X	X	X	X
10	X	X	X	X

Similarly, the expressions for JC and KC we obtain as:

$JC = A'B'$ and

$KC = B'.$

A Karnaugh map has been also prepared below for output y and the Boolean expression for y is obtained as:

$Y = Bx' + BC + Ax$

For Y

AB \ Cx	00	01	11	10
00	0	0	0	0
01	1	0	1	1
11	X	X	X	X
10	0	1	X	X

The circuit diagram of the desired sequential logic network is shown in the below figure:

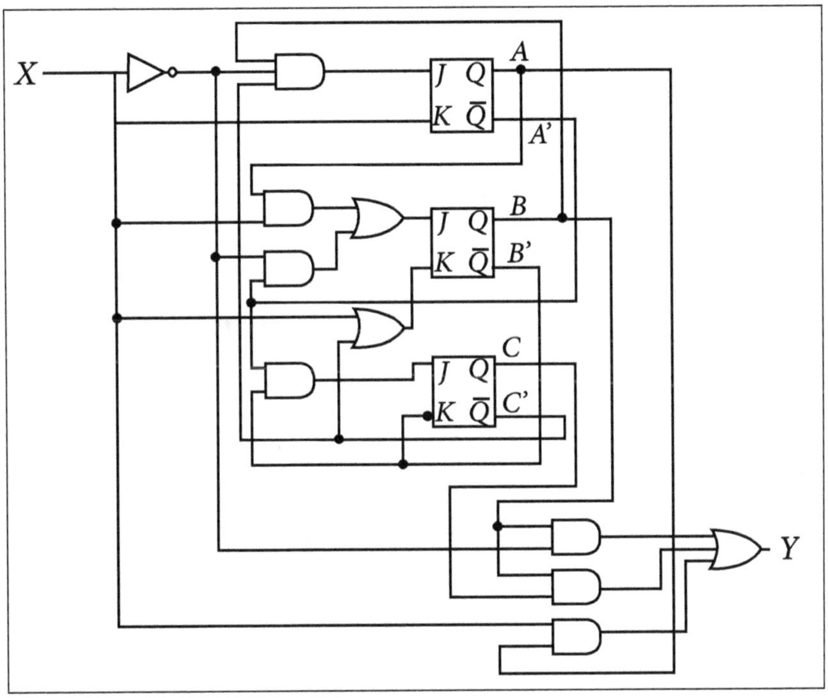

The actual design procedure is illustrated with the help of few examples:

- Type of FF to be included in the design generally depends on what is available to the designer. Many digital systems are made with JK FFs because these are the most usually available these days.
- When many types of FFs are available, it is advisable to use D-FFs for applications requiring transfer of data (shift-registers): T-type for applications involving complementation (counters) and J-K type for general applications.
- N FFs will give us up to 2" distinct states.
- Proper care should be taken to connect interpretation of circuit specifications.
- The information from the truth-table is transferred to K-maps and minimized equations desired. Combinational circuit design then follows its implementation.

Problems

1. Design a clocked sequential circuit, using 3 FFs A. B and C, one (I/P) x and one (O/P) y. The state diagram is shown in the below figure using J-K-FFs and treating unused states as don't care condition. Analyses the final circuit to be self-correcting.

Solution:

Given:

Using the minimized equations will be as follows:

$J_A = \overline{B} \cdot X;$ $K_A = 1$

$J_B = A + \overline{C} \cdot \overline{X};$ $K_B = \overline{C} \cdot X + C \cdot \overline{X}$

$J_C = A \cdot X + \overline{A} \cdot \overline{B} \cdot \overline{X};$ $K_C = X$

$y = \overline{A} \cdot X$

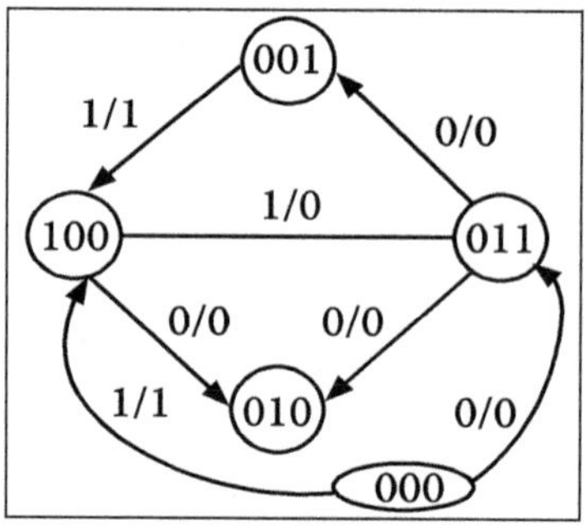

State Diagram.

To implement these equations, suitable combinational circuits can be designed.

2. Design a sequential circuit, using T-FFs, to examine a train of binary digits. The output must be '1', when the last 4 digits received in the train, are of the form 0010.

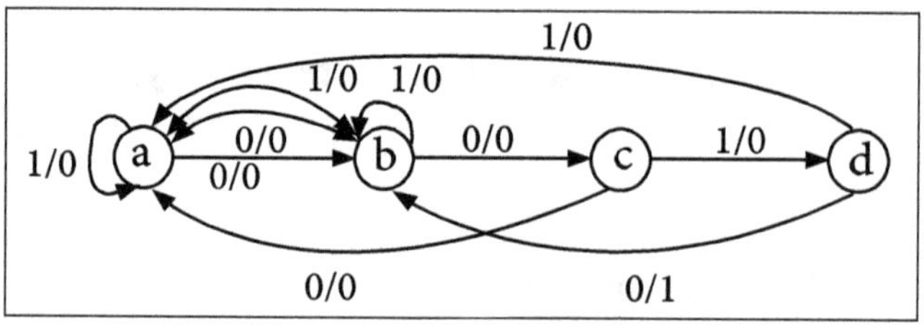

State Diagram.

State Table can be given as:

Present state		Next State				O/P	
		X=0		X=1		X=0	X=1
a	00	b	01	a	00	0	0
b	01	c	10	a	01	0	0
c	10	a	00	d	11	0	0
d	11	b	01	a	00	1	0

Excitation Table

Present state	(I/P)	Next state	O/P	FF/ I/PS	
				TA	TS
0 0	0	0 1	0	0	1
0 0	1	0 0	0	0	0
0 1	0	1 0	0	1	1
0 1	1	0 0	0	0	0
1 0	0	0 0	0	1	0
1 0	1	1 1	0	0	1
1 1	0	0 1	1	1	0
1 1	1	0 0	0	0	0

T-FF Excitation Table

Q_1	Q_t+1	T
0	0	0
0	1	1
1	0	1
1	1	0

K-maps for T_A, T_B and Y

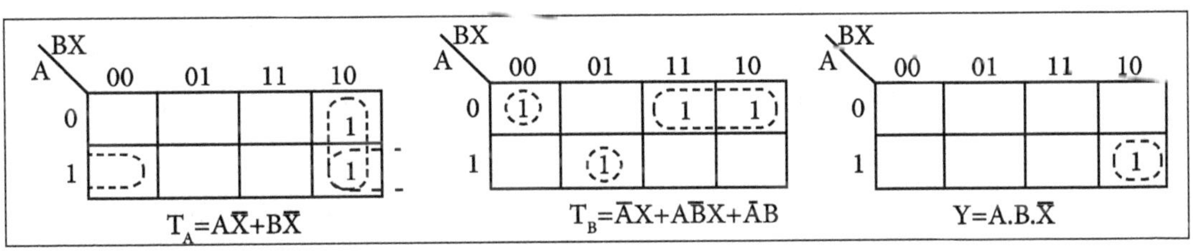

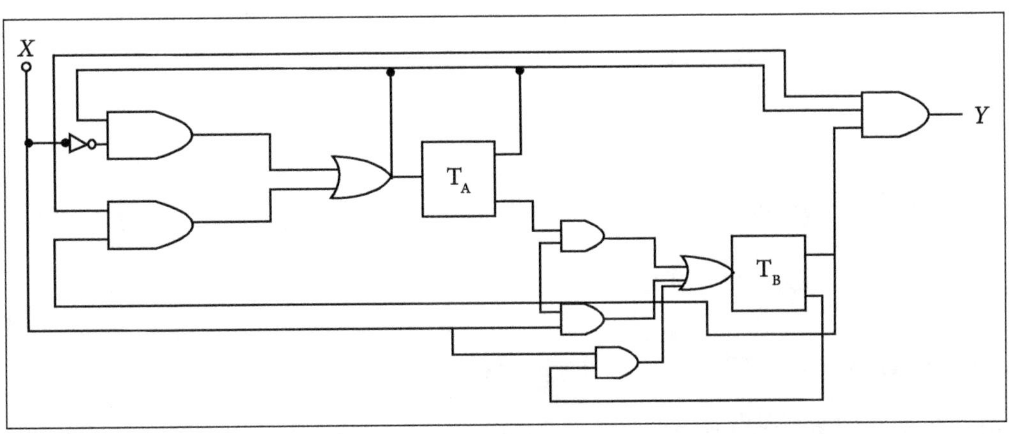

Implementing Combinational Circuit with 2-T-FF'S.

6.4 Realization of Circuits using Various Flip-Flops

We already know that a D flip-flop can be formed from a J-K flip-flop: simply tie the set input J to the data input D and the reset input K to D'.

In similar fashion, a T flip-flop can be derived from a J-K flip-flop by connecting both the J and K inputs to T.

Choosing a Flip-Flop Type

R-S Latch: The level-sensitive R-S latch has limited utility as a stand-alone building block for holding state. However, it is used as a component in implementing master/slave or edge-triggered flip-flops and it serves some special functions for debouncing switches.

J-K Flip-flops Venus D Flip-Flops: In real designs, the most frequently encountered flip-flops are the J-K and the D. The combinational logic associated with determining the next state will often require the fewest gates when the next state function is implemented with J-K flip-flops.

However, each J-K flip-flop requires two inputs and this could lead to more complex wiring. If the goal is to minimize wires, D flip-flops are attractive. In VLSI technologies, where the size of the wiring area is typically a greater concern than gate area, D flip-flops are used almost universally and are usually implemented as cross-coupled inverters.

T Flip-flops: T flip-flops are rarely available in packaged logic because they are so easily formed from J-K flip-flops. From the viewpoint of conceptual design, however, T flip-flops turn out to be good building blocks for special circuits called counters. These are clocked sequential systems that sequence through a periodic series of states, such as the binary numbers 0000 through 1111.

If Q is the current state and Q+ is the next state, the equations for the four flip-flop types are:

$$\text{R}-\text{S latch}: Q_+ = S + \overline{R}Q$$

$$\text{D flip}-\text{flop}: Q_+ = D$$

$$\text{J}-\text{K flip}-\text{flop}: Q_+ = \sqrt{Q} + \overline{K}Q$$

$$\text{T flip}-\text{flop}: Q_+ = T\overline{Q} + \overline{T}Q$$

Conversion of One Flip-Flop Type to another

Any flip-flop can be implemented as combinational logic for the next state function in conjunction with a flip-flop of another type.

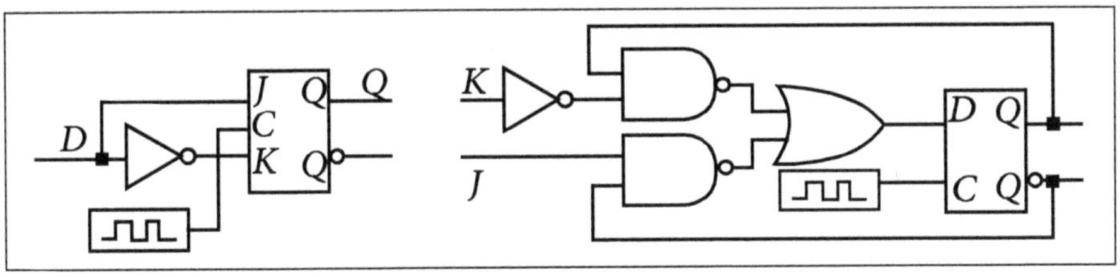

D Flip-flop implemented by J-K Flip-flop.

As an example, the above figure shows how to implement a D flip-flop with a J-K flip-flop and, correspondingly, a J-K flip-flop with a D flip-flop.

Let us consider the leftmost circuit. If D is 1, we place the J-K flip-flop in its set input (J = 1, K = 0). If D is 0, J-K's inputs are configured for reset (J = 0, K = 1). In the case of the rightmost circuit, the D flip-flop's input is driven with logic that implements the characteristic equation for the J-K flip-flop, namely $\sqrt{Q}+\overline{K}Q$.

General Procedure we can follow a general procedure to map among the different kinds of flip-flops. It is based on the concept of an excitation table, that is, a table that lists all possible state transitions and the values of the flip-flop inputs that cause a given transition to take place.

Excitation Tables for R-S, J-K, T and D Flip-Flops

Q	Q'	R	S	J	K	T	D
0	0	X	0	0	X	0	0
0	1	0	1	1	X	1	1
1	0	1	0	X	1	1	0
1	1	0	X	X	0	0	1

The above gives excitation tables for R-S, T and D flip-flops. If the current state is 0 and the next state is to be 0 too, then the first row of the table describes the flip-flop input to cause that state transition to take place. If an R-S latch is being used, it does not matter what value is placed on R as long as S is left unasserted. R = 0, S = 0 holds the current state at 0; R = 1, S = 0 resets the state to 0. The effect is the same.

If we are using a J-K flip-flop, the transition from 0 to 0 is accomplished by ensuring that J is left unasserted. The value of K does not matter. If J = 0, K = 0, the current state is held at 0; if J = 0, K = 1, the state is reset to 0.

If we are using a T flip-flop, the transition does not change the current state, so the input should be 0. If a D flip-flop is used, we set the input to the desired next state, which is 0 in this case. The same kind of analysis can be applied to complete the excitation table for the three other cases.

A flip-flop's next state function can be written as a K-map.

For example, the next state K-map for the D flip-flop is shown in the below figure:

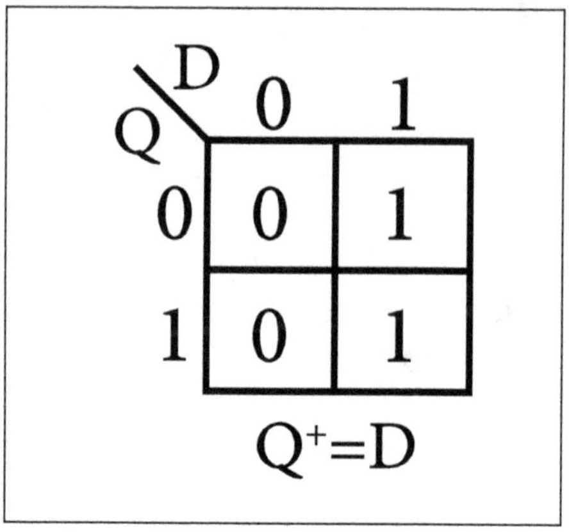

(a) K-map for D Flip-flop and Characteristic equation.

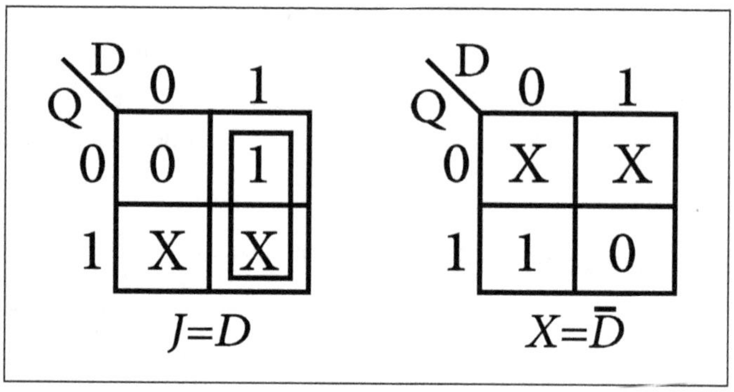

(b) K-maps for J and K input.

To realize a D flip-flop in terms of a J-K flip-flop, we simply remap the state transitions implied by the D flip-flop's K-map into equations for the J and K inputs. In other words, we express J and K as functions of the current state and D.

The procedure works as follows:

First we draw K-maps for J and K, as in above figure (a) & (b). Then we fill them in the following manner.

When D = 0 and Q = 0, the next state is 0. The excitation table tells us that the inputs to J and K should be 0 and X, respectively, if we desire a 0-to-0 transition. These values are placed into corresponding entries of the J and K K-maps.

The inputs D = 0, Q = 1 lead to a next state of 0. This is a 1-to-0 transition and J and K should be X and 1, respectively. For D = 1, Q = 0, the transition is from 0 to 1 and J must be 1 and K should be X. The final transition, D = 1, Q = 1, is from 1 to 1 and J and K are X and O. A quick look at the K-maps confirms that J = D and K = D'.

The implementation of a J-K flip- flop by a D flip-flop follows the same procedure. We start with a K-map to describe the next state in terms of the three variables J, K and the current state Q. To obtain the transition from 0 to 0 or 1 to 0 requires that D be 0; similarly, D must be 1 to implement a 0-to-1 or 1-to-1 transition. In other words, the function for D is identical to the next state.

The equation for D can be read directly from the next state K-map for the J-K flip-flop.

$$D=\sqrt{Q}+\overline{K}Q$$

This K-map is shown in the below figure:

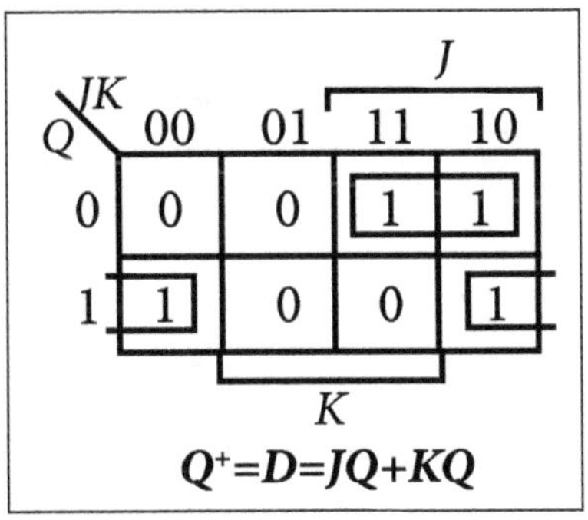

Problems

1. Let us consider that the following sequence is to be realized by a counter consisting of 3 JK Flip flops.

A_1	0	0	0	0	1	1	0
A_2	0	1	1	0	0	1	0
A_3	0	1	0	1	1	0	0

Here we will design the counter.

Solution:

Given:

A_1	0	0	0	0	1	1	0
A_2	0	1	1	0	0	1	0
A_3	0	1	0	1	1	0	0

Truth Table

Present State			Next State			Flip-Flop Inputs					
A_1	A_2	A_3	A_1	A_2	A_3	JA_1	KA_1	JA_2	KA_2	JA_3	KA_3
0	0	0	0	1	1	0	X	1	X	1	X
0	1	1	0	1	0	0	X	X	0	X	1

0	1	0	0	0	1	0	X	X	1	1	X
0	0	1	1	0	1	1	X	0	X	X	0
1	0	1	1	1	0	X	0	1	X	X	1
1	1	0	0	0	0	X	1	X	1	0	X

K – Map

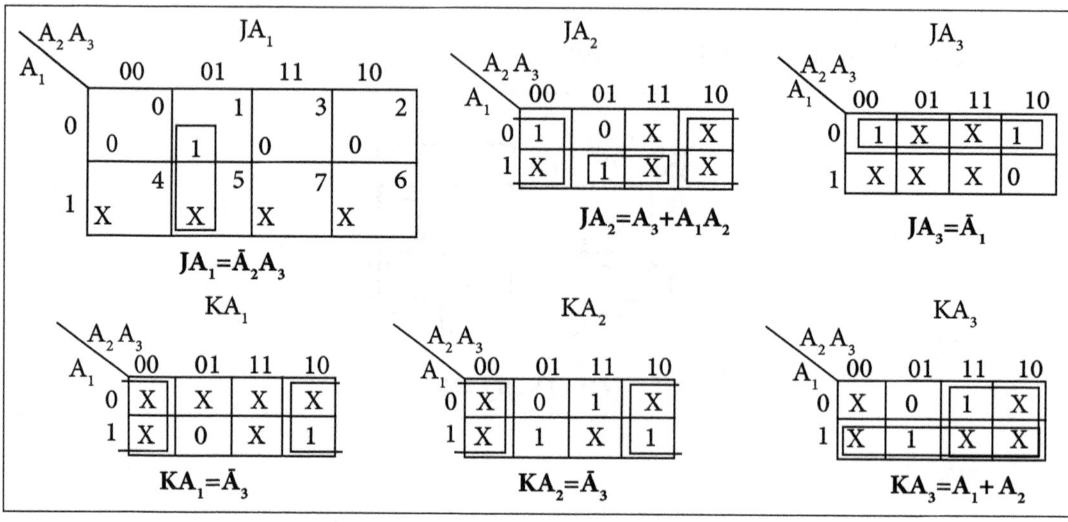

State Diagram

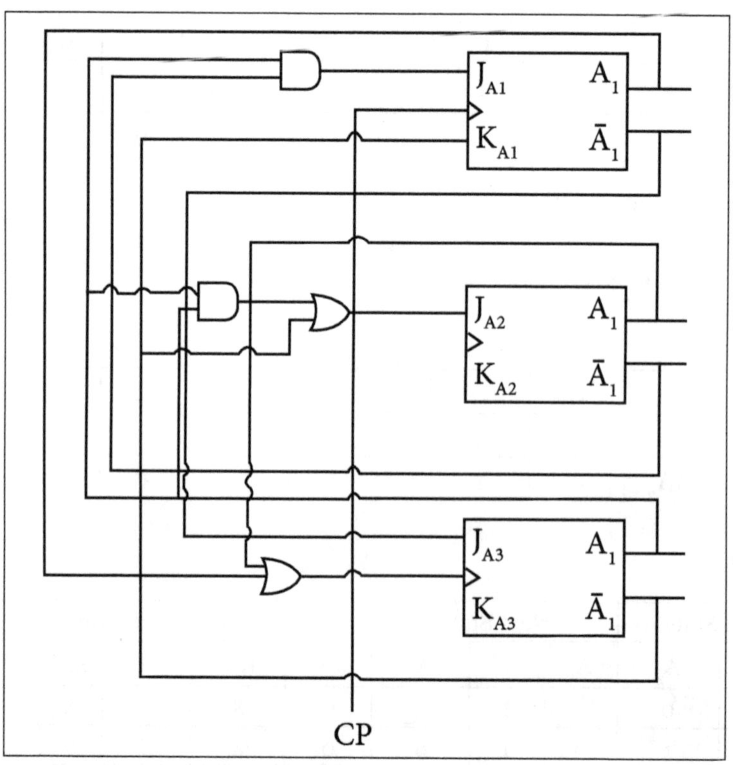

2. Let us design the synchronous sequenced circuit using JK flip-flop for 0, 1, 2, 4, 5, 6.

Solution:

Given:

0→1→2→4→5→6

State Diagram

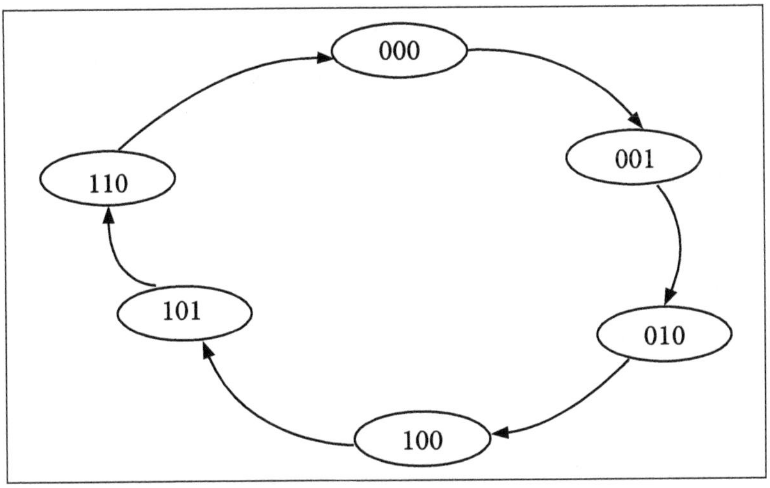

Present State			Next State		
A_t	B_t	C_t	A_{t+1}	B_{t+1}	C_{t+1}
0	0	0	0	0	1
0	0	1	0	1	0
0	1	0	1	0	0
1	0	0	1	0	1
1	0	1	1	1	0
1	1	0	0	0	0

6.5 Mealy to Moore Conversion and Vice-Versa

Finite automata may have outputs corresponding to each transition. There are two types of finite state machines that generate output. They are:

- Mealy Machine
- Moore machine

Mealy and Moore Model

If the output function depends on the present state and present inputs, it is called the Mealy model, named after G.H. Mealy, a pioneer in this field. If the output is associated only with the present state, it is called the Moore model, named after another pioneer E.F. Moore. The counters are clearly Moore machines as the output depends only on the states of the Flip-flops. Likewise, a sequence detector is also a Moore machine. Serial adder is an example of a Mealy machine as each one of its states is reached producing a o or I output depending on the starting state and the value of the inputs. Fundamental mode asynchronous sequential machines are clearly Mealy machines as the outputs depend on the input column apart from the internal state.

Mealy Machine

A Mealy Machine is an FSM whose output depends on the present state as well as the present input.

It can be described by a 6 tuple $(Q, \sum, O, \delta, X, q_o)$.

Where,

- Q is a finite set of states.
- $\sum$ is a finite set of symbols called the input alphabet.
- O is a finite set of symbols called the output alphabet.
- δ is the input transition function where $\delta : Q \times \sum \rightarrow Q$.
- X is the output transition function where $X : Q \times \sum \rightarrow Q$.
- q_o, is the initial state from where any input is processed $(q_o \varepsilon Q)$.

The state table of a Mealy Machine is shown below:

Present state	Input =0		Input =2	
	State	Output	State	Output
$\rightarrow$a	b	X_1	c	X_1
b	b	X_2	d	X_3
c	d	X_3	c	X_1
d	d	X_3	d	X_2

The state diagram of the above Mealy Machine is shown:

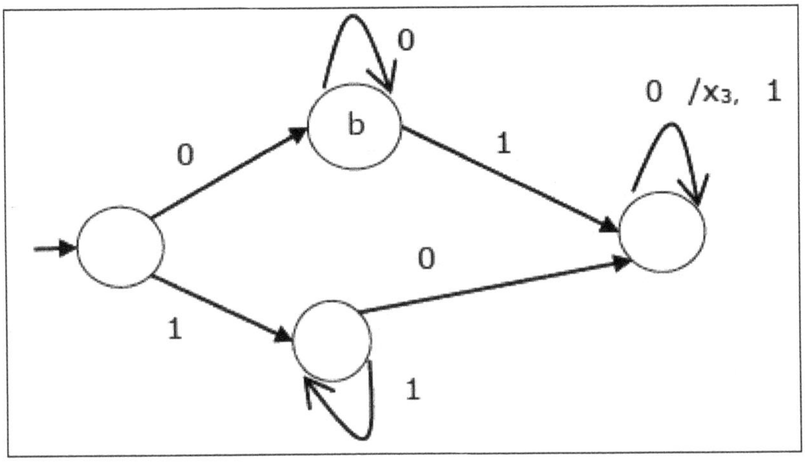

Mealy to Moore conversion

Let us use the familiar serial adder as an example to illustrate the procedure. The state diagram and the state table of the synchronous serial adder are given in the below figure respectively. The state P is reached from the state Q on the application of the inputs AB = 00 and, in the process, the machine produces an output Z = 1 indicated on the arc as 00/1.

Also it is important that the machine produces output Z = 0 in another transition to P. This transition is indicated as a self-loop around P on inputs AB = 00/0. For AB = 01 or 10 while in P, the machine produces an output I and remains in the same state P.

Since, the transition to state P is associated with both outputs, 0 and 1, we split the state into two equivalent states and call them P_0 and P_1. Every transition to P with an output 0 will be directed to P_0 and every transition to P with an output 1 will be directed to P_1. For similar reasons, state Q is also split into Q_0 and Q_i. The modified state table and the state diagram with four states are shown in the following respectively.

Clearly, the modified machine of the below figure, which is in the form of a Moore model, has to be equivalent in behavior to the Mealy model of the given below figure.

We make this statement as the two machines have to give the same output sequence for any arbitrary input sequence. The reader may make two observations.

- If the Mealy machine has K states, the equivalent Moore machine will have at most 2K states if the output is a binary variable.
- There is no power-on state, unless specifically defined. A special state may be introduced if the user wants one.
- In a Mealy machine, the next state and output are given at the same place separated by a comma while a Moore machine has an exclusive output column.

Algorithm

- Input - Mealy Machine
- Output - Moore Machine

Step 1: Calculate the number of different outputs for each state (Q_i) that are available in the state table of the Mealy machine.

Step 2: If all the outputs of Qi are same, copy state Q. If it has n distinct outputs, break Q_i into n states as Qin where n = 0, 1, 2……

Step 3: If the output of the initial state is 1, insert a new initial state at the beginning which gives 0 outputs.

Example:

Let us consider the following Mealy Machine.

Present state	Next State			
	a = 0		a =1	
	Next State	Output	Next State	Output
→a	d	0	b	1
b	a	1	d	0
c	c	1	c	0
d	d	0	a	1

Here, states 'a' and 'd' give only 1 and 0 outputs respectively, so we retain states 'a' and 'd'. But states 'b' and 'c' produce different outputs (1 and 0). So, we divide b into b_0, b_1 and c into c_0, c_1.

Present State	Next State		Output
	a =0	a = 1	
→a	d	b1	1
bo	a	d	0
b1	a	d	1
Co	C1	Co	0
C1	C1	Co	1
d	bo	a	0

Moore to Mealy Conversion

It is amazingly simple to convert a Moore machine into a Mealy machine.

Let some state S_i of a given Moore machine be associated with an output Z_i. We have to associate the output Z_i wherever Si occurs as the next state by scanning all the input columns. This is indicated by replacing all Si by S_i, Z_i. We then call the state S_i as Z homogeneous. All states of a Moore machine are Z homogeneous.

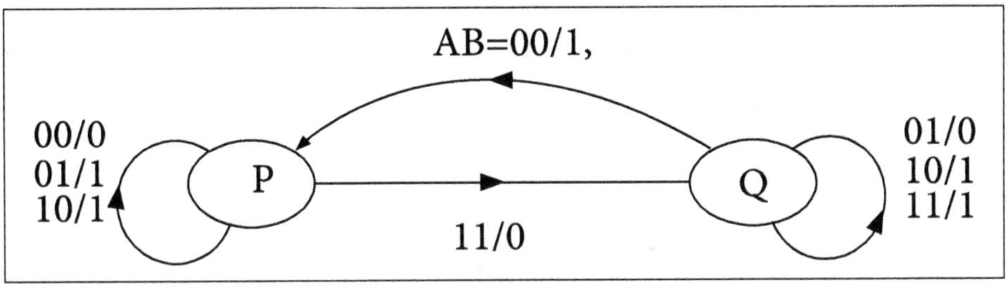

(a) State diagram of a serial adder as a Mealy machine.

PS	NS , Z Inputs AB			
	00	01	11	10
P	P, 0	P, 1	Q, 0	P, 1
Q	P, 1	Q, 0	Q, 1	Q, 0

(b) Mealy state table.

	NS , AB				Z
PS	00	01	11	10	
P_0	P_0	P_1	Q_0	P_1	0
P_1	P_0	P_1	Q_0	P_1	1
Q_0	P_1	Q_0	Q_1	Q_0	0
Q_1	P_1	Q_0	Q_1	Q_0	1

(c) Moore state table.

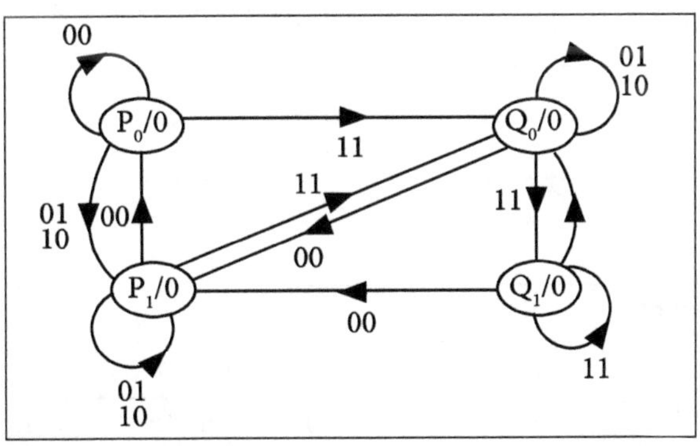

(d) State diagram of a serial adder as a Moore machine.

Algorithm

- Input - Moore Machine
- Output - Mealy Machine

Step 1: Take a blank Mealy Machine transition table format.

Step 2: Copy all the Moore Machine transition states into this table format.

Step 3: Check the present states and their corresponding outputs in the Moore Machine state table; if for a state Qi output is m, copy it into the output columns of the Mealy Machine state table wherever Qi appears in the next state.

Example:

Let us consider the following Moore machine.

Present State	Next State		Output
	a = 0	a = 1	
→a	d	b	1
b	a	d	0
c	c	c	0
d	b	a	1

Now we apply above mentioned to convert it to Mealy Machine.

Steps 1 & 2

Present State	Next State			
	a = 0		a = 1	
	State	Output	State	Output
→a	d		b	
b	a		d	
c	c		c	
d	b		a	

Step 3

Present State	Next State			
	a = 0		a = 1	
	State	Output	State	Output
=>a	d	1	b	0
b	a	1	d	1
c	c	0	c	0
d	b	0	a	1

Problem

1. Here let us draw the state transition diagram of a sequence detector circuit that detects '1010' from input data stream using Moore model, Mealy model.

Solution:

Given: Input data – 1010

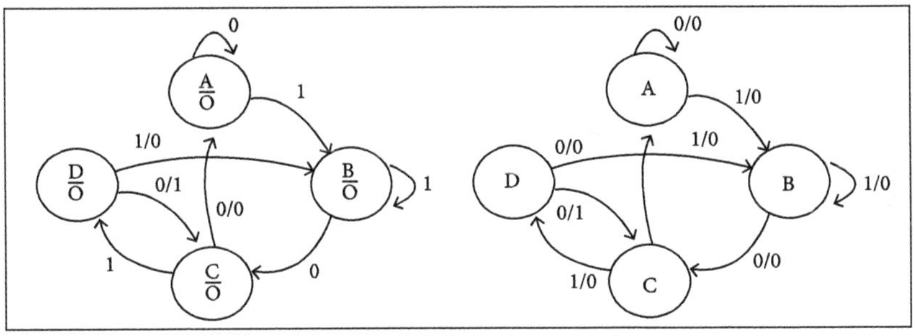

Moore model Mealy model.

State Table

PS	NS		Y	
	X = 0	X = 1	X = 0	X = 1
A	A	B	0	0
B	C	B	0	0
C	A	D	0	0
D	C	B	1	0

Permissions

All chapters in this book are published with permission under the Creative Commons Attribution Share Alike License or equivalent. Every chapter published in this book has been scrutinized by our experts. Their significance has been extensively debated. The topics covered herein carry significant information for a comprehensive understanding. They may even be implemented as practical applications or may be referred to as a beginning point for further studies.

We would like to thank the editorial team for lending their expertise to make the book truly unique. They have played a crucial role in the development of this book. Without their invaluable contributions this book wouldn't have been possible. They have made vital efforts to compile up to date information on the varied aspects of this subject to make this book a valuable addition to the collection of many professionals and students.

This book was conceptualized with the vision of imparting up-to-date and integrated information in this field. To ensure the same, a matchless editorial board was set up. Every individual on the board went through rigorous rounds of assessment to prove their worth. After which they invested a large part of their time researching and compiling the most relevant data for our readers.

The editorial board has been involved in producing this book since its inception. They have spent rigorous hours researching and exploring the diverse topics which have resulted in the successful publishing of this book. They have passed on their knowledge of decades through this book. To expedite this challenging task, the publisher supported the team at every step. A small team of assistant editors was also appointed to further simplify the editing procedure and attain best results for the readers.

Apart from the editorial board, the designing team has also invested a significant amount of their time in understanding the subject and creating the most relevant covers. They scrutinized every image to scout for the most suitable representation of the subject and create an appropriate cover for the book.

The publishing team has been an ardent support to the editorial, designing and production team. Their endless efforts to recruit the best for this project, has resulted in the accomplishment of this book. They are a veteran in the field of academics and their pool of knowledge is as vast as their experience in printing. Their expertise and guidance has proved useful at every step. Their uncompromising quality standards have made this book an exceptional effort. Their encouragement from time to time has been an inspiration for everyone.

The publisher and the editorial board hope that this book will prove to be a valuable piece of knowledge for students, practitioners and scholars across the globe.

Index

Printed in the USA
CPSIA information can be obtained
at www.ICGtesting.com
LVHW082210201023
761596LV00009B/1654

9 798888 360439